A Nature Study Guide

W. S. Furneaux

Copyright Notice

Published by Cedar Lake Classics

TABLE OF CONTENTS

Preface

The value of nature study as a means of training children to observe and investigate is now fully recognized by the majority of our best teachers, with the result that the careful study of natural things and phenomena takes a very prominent place in the school curriculum; and the object of this little guide is to assist the teacher in his attempts to obtain for the children the maximum benefit of the thoughtful observations of their physical environment.

The purpose of the book is not to supply the teacher with information on all the various aspects of Nature, for an attempt to attain this end in a single volume would necessarily result in a most scrappy and unsatisfactory summary of Nature's works.

The aim is rather to lead the teacher to the best methods of treating his subjects, and to supply him with such practical suggestions as will help him in providing and maintaining a suitable supply of material for both occasional and continuous observations.

Thus, while a certain amount of information is given with the object of calling attention to various things of special interest, and to phenomena that are not always understood, the space is devoted mainly to the treatment of nature lessons within the school building, to seasonal studies outdoors, and to the preparation and management of valuable aids to the study of Nature, such as the aquarium, the vivarium, the school garden, and the school museum.

Although but little space is devoted to the descriptions of natural objects, it is hoped that the numerous photographs and other illustrations will enable the reader to identify the majority of the things named.

W. S. Furneaux
London, 1911.

— 1 —

Introduction to Nature Study

The study of Nature now takes a prominent place in the curriculum of many schools, and while many teachers regard it as being a valuable aid in the training of infants and junior scholars, others have fully recognized its usefulness as a study for children throughout the whole period of their school life.

But we must, at the outset, state precisely what we mean by the term 'Nature Study.' It is the careful and thoughtful observation of natural objects and natural phenomena by the children, under the guidance of the teacher—a process of research on the part of the children by means of which natural objects and phenomena acquire meaning.

It will be clearly seen from the above definition that we have nothing whatever to do with the old type of object lesson in which information acquired by the teacher is imparted to the class, not even if such a lesson is illustrated by the exhibition of the object to question, as well as by the best of pictures or diagrams.

Such a lesson is merely a lesson of information, in which the children gain second-hand knowledge; and the acquisition of the facts given is only a matter of memory, unaccompanied by those important mental processes which assist in the development of the growing mind. The true nature study lesson is one in which each child closely observes an object placed before him, or studies a phenomenon that presents itself to him at the time, and in which he is encouraged to form his own conclusions, and to realize, as far as possible, the true nature of the thing seen.

Thus nature study, as we are to understand it, is to be looked upon rather as a method than as a subject. It is, with the teacher, an effort to bring the children in direct contact with things, to cultivate the habit of careful observation and discrimination, to create a living interest in the surroundings, and to encourage independent thought. It teaches the child not only to see, but to recognize; and it produces a habit of sensory alertness at a period during which the mind is particularly plastic and impressionable.

There is a vast difference between nature knowledge and nature study. The former simply denotes facts acquired, while the latter is rather a spirit of inquiry and research by which natural objects and phenomena arouse a living interest and encourage investigation. In the latter case the work of the teacher is not to give information, but rather to stimulate the children to observe and discriminate for themselves, and to form their own conclusions.

Of course, in the case of young children, the ideas formed and the conclusions framed will always be more or less vague and imperfect; but since these ideas and conclusions are the result of the children's own efforts, they are of far more value than the clearer conceptions imposed by the teacher on a class that is merely passively receptive.

The value of nature study as a means of training children can hardly be overestimated. The habit of close and thoughtful observation that it cultivates will not only have a great influence on them during their period of school life, but will also assist them in their future careers. It will help them to see and understand various natural objects and the phenomena associated with them that would otherwise remain practically unnoticed, and will have a very great influence in determining their tastes and pursuits.

This cultivated habit of closely observing natural objects and phenomena will give the child a practical grasp of the whole physical world, enabling him to recognize all things and occurrences as a set of conditions that form his own environment. It will produce a keenness of the senses and precision of observation that, coupled with an appreciative interest in the surroundings and a natural inquisitiveness concerning things in general, will put him in a much better position to carry out the work demanded of him in his future career with initiative, self-reliance, and a productive method.

The training which nature study gives not only causes the child to see with the mind as well as with the eye, but teaches him to observe with a purpose; and the mental discipline it enforces provides a splendid foundation for the future study of the experimental sciences. A good systematic course of nature study will also lead to neatness, accuracy and dexterity in all work undertaken, and do much towards the cultivation of patience and perseverance in the worker.

If the effect of a good course of nature study is to produce in the child all that we claim for it, it is clear that the training must have more or less influence in connection with the teaching of all school subjects. But some of these subjects are so closely allied to this study that they should be worked hand in hand with the latter. Thus the drawing lessons and the clay modeling exercises may be continuations of the study of the natural objects examined, and the teaching of geography may be conducted as an extension of the outdoor observations of natural objects and phenomena.

Then again, a very large proportion of our best literature teems with references to natural objects and phenomena, and thus the study of Nature enables us to understand and enjoy much that would otherwise be meaningless or vague.

There is yet another aspect of the subject well worth consideration. Nature study is certainly of great value as an aid towards the culture of aesthetic tastes, and many of our best teachers further recognize in it a powerful aid in moral training. It cultivates the judgment and the imagination, and thus leads to such thoughtful and intelligent observation that the child not only becomes acquainted with the facts of Nature, but sees and appreciates her beauties and realizes her wonders. This appreciation of the beauties and wonders of Nature leads to a sympathy with all living things, thus correcting the natural tendency to destructiveness; and it also tends to create a broad human sympathy. No study so thoroughly arouses the aesthetic and emotional elements of a child's character, and no school study can do more to brighten the lives of the children.

—2—

Nature Lessons

A. Choice of Subjects

In selecting subjects and in making out schemes for a course of nature study it is absolutely essential that we follow the course of the seasons, so that each of the studies may be made direct from the fresh or living material, and the various natural phenomena engage attention at the times of their occurrence.

The work should not consist of a series of set lessons, rigidly defined as to time and character, with no logical connection between them; but of a carefully prepared scheme of observations, drawn up in perfect accordance with the succession of the seasons, and so arranged that each portion naturally evolves itself from that which precedes it.

Such a scheme, while systematic from beginning to end, must not be too rigid. The very seasons on which it is based are themselves so variable that it would be very unwise to fix the date on which each portion of the work is to be done; and it would be equally unwise to attempt to decide how much should be done in a given period. The most experienced teacher is unable to foresee the many difficulties which may arise in the minds of the children—difficulties that should, as far as possible, be cleared away before the next steps are taken in hand; nor can he foresee the occasional disappointments that sometimes occur in connection with the collection of material for his work, and, on the other hand, the unexpected wealth of material that will now and then fall in his way.

Furthermore, whatever may have been the care bestowed on the preparation of a scheme of nature observations, in the hands of a thoughtful teacher, new ideas and developments are sure to present themselves; and, for this reason alone, the teacher should have perfect liberty to adjust the work as it proceeds, rather than feel himself compelled to follow a stereotyped course in which his own initiative and that of the children are more or less restrained.

Again, the work laid out should never be excessive. The value of the work done is not to be gauged by the number and variety of subjects compressed into the scheme, but rather by the thoroughness of that which has been done. And, as regards the nature of the work introduced, it is probable that nothing is more effectual in the training of young minds than the continued observations of a progressive series of events such as those exhibited in the development of seedlings under varying conditions, in the varied aspects of trees at the different seasons of the year, and in the life history of an insect or other creature traced from the egg to the adult or perfect stage.

Many of the subjects that may well form part of a nature study course are such that they can only be successfully dealt with on certain rare or special occasions. Thus, we take the opportunity of studying the snowstorm while such a storm is in progress; and, similarly, the thunderstorm and other occasional atmospheric disturbances at the times when they occur. We also call attention to the differences between stars and planets during a period when one or more of the latter are conspicuous in the sky. The migrations of birds are studied at those seasons when the movements are taking place; and the hibernations of various animals during the autumn, when they are making preparations for the long winter sleep, and during the winter itself, when they may be observed in their snug hiding places. In short, as previously laid down, every subject must be taken in its proper season, so that the whole scheme is in perfect harmony with the daily experiences of the children.

It is the writer's experience that most teachers find a greater difficulty in the selection of suitable subjects from the animal than from the vegetable world. This is partly due to the fact that common British animals are not so generally studied as are the common flowers and trees. The lower animals—the invertebrates—are especially neglected on account of the general aversion towards creeping things.

The old-fashioned, so-called nature lesson, illustrated only by a picture and, perhaps, a fragment of skin, hoof or horn, is of very little educational value. The cardinal feature of animal life is motion; and if the children have not the opportunity of observing the interesting habits of the animal in question, and of working out the striking relation that exists between the habits and the structure, the charm and value of the lesson are lost.

Seeing that the object of the nature lesson is not to supply information, but to encourage independent observation and discrimination, it is clear that one animal is practically as useful for the purpose as any other; and, therefore, there is no reason why, as a rule, the lesson should not be based on some form of animal life that can be conveniently studied within the schoolroom, or that may be observed in the neighborhood of the building.

Of course, we do not mean that no information should ever be given on foreign animals and on those British species which can seldom or never be seen alive by the children. Such information may often be extremely useful in connection with the teaching of geography—a subject that is very closely allied to nature study. But the information so given should not constitute a set lesson in itself, for the mere presentation of facts by the teacher is not of sufficient importance to demand much time, and a lesson partaking of the character referred to is entirely foreign to the spirit of nature study.

An enthusiastic student of Nature will soon discover that there is a wonderful wealth of animal forms among British species which are eminently suitable for study by children; for, in addition to our familiar mammals and birds, we have many interesting fishes in our ponds and streams, a few amphibians (frogs, toads, and newts) with exceptionally interesting life histories, harmless reptiles, and many invertebrates, such as insects, spiders, snails, earthworms, etc., the majority of which may be easily kept in captivity for constant observation, or studied in their natural habitats in the neighborhood of the school.

The scheme of nature study set out on future pages will, we hope, give many useful suggestions to the teacher; and the various hints on the treatment of creatures that may be kept in captivity either in the school garden or in the schoolroom itself will enable him to maintain a wealth of living material for close and systematic observation.

The collection of specimens for the study of Nature need not, and should not, devolve entirely upon the teacher. Let the children once get an insight into the wonders of natural objects around them, and they will always be on the alert for new sources of delight, with the result, especially in the case of schools in the country or on the outskirts of towns, that more than sufficient material will generally be forthcoming for the nature study work.

It is well to encourage this propensity for the collection of natural objects on the part of the children, providing it is properly directed. Care should be taken to secure that the children do not develop into mere collectors of material without discrimination as to the usefulness or otherwise of the specimens acquired. Their labors in this direction should be so controlled that they bring in only such material as is necessary in the working of the nature study scheme of the school, together with those objects concerning which they desire to gain information.

This latter point is one of considerable importance, for it is the duty of the teacher to encourage the natural curiosity of the children under his charge; and it will be well, now and then, to devote a little time to pleasant chats on their observations and specimens, even though they do not fall within the range of the course planned for the schoolwork. Such chats will not only be a source of much delight, but will also be a wonderful stimulus for keen and thoughtful observation in the future.

B. Nature Lessons

Let us now pass on to consider the matters which relate more directly to the regular lessons that make up the nature study course of the school, leaving, for the present, those occasional observations which, though forming an important part of the scheme in operation, do not require set times and periods.

In accordance with the old plan which insisted on some kind of "introduction" to the lesson, the question is often asked: "How shall I introduce this lesson?"

A nature lesson requires no formal, spoken introduction by the teacher. Set the object to be studied before the class, and let the observations commence at once. The commonest form of introduction to a lesson is, perhaps, a series of questions put by the teacher with the object of encouraging the children to guess what he is going to talk about. This is, of course, an absolute waste of time; and even where the lesson naturally evolves itself from a preceding one, and it is necessary for the children to see the connection between the present subject and the last, this connection is often best seen after the present lesson has been practically concluded and the relation between the two should be worked out by the children, and not by the teacher.

Certainly one of the best ways in which to start a nature lesson is to place the object of study before the children, and then tell them to observe carefully, the teacher himself being careful to allow ample time for a very thorough inspection of the specimens.

Some would insist that this introductory observation of the specimens should be perfectly silent, the view being entertained that children should not be allowed to talk in school—that the discipline of the school—the power of the teacher over his class—would suffer if such liberties were allowed; but if the tone of the school is what it should be, the dignity and power of the teacher will lose nothing from the permission given to the children to exchange observations and thoughts with one another. It is astonishing, too, to observe how children, left for a time to themselves, can help each other in the discovery of facts and in the solving of little problems, to say nothing of the increased interest in their subject brought about by communication of their discoveries and ideas.

Of course the observations of the children, under these conditions, will be carried on regardless of any definite order; and the ideas framed may often be somewhat confused and incorrect. But the children should have the first opportunity of seeing, and the first opportunity of investigating. Where necessary, the teacher may, by an occasional remark, direct the observations into some desired order, and any confusion of ideas may afterward be corrected.

After the interest of the class has been thoroughly aroused by a preliminary observation of this kind, the teacher demands the attention of the children and, by a carefully planned series of questions, discovers what observations have been made, and draws attention to other points which should have been seen.

Further questions will be asked with the object of encouraging the children to think out simple problems with regard to the habits and mode of growth of the thing before them, and to work out the uses and functions of its various parts.

Throughout the whole lesson the teacher should be careful to do nothing for the children which they can do for themselves—to tell them nothing which they themselves can discover, and to offer no explanation where it is possible for them to solve the matter themselves. He should give the required assistance only where the children fail after every possible encouragement has been given, and remember that the inability of the children to observe certain points of structure and to think out the problems involved is often due to more or less impatience on the part of the teacher, resulting from his desire to get on with his subject in order that the lesson may be completed within a given time.

This latter error is a grave one. It is quite right that a teacher should carefully plan out his work, and form some kind of estimate as to what he is likely to do in the time at his disposal, but he should never attempt to adjust the progress of the lesson in order to make it coincide with the time. It matters not whether a lesson is completed according to the plan laid out, but it is most important that the work done is done thoroughly.

For this reason the teacher has a right to demand the fullest liberty in dealing with his subject. He never knows what difficulties will arise during the progress of the lesson. Many unexpected points of interest will frequently present themselves. Occasionally it will happen that a topic, concerning which the teacher anticipates a difficulty, turns out to be less formidable than was supposed. Hence he should have full power to expand or omit any portion of the work previously planned, and even to change the order originally proposed, when he is of opinion that by so doing he can make his work more productive.

We have spoken of the importance of careful questioning on the part of the teacher, but we must note that the children should be allowed and, indeed, strongly encouraged to put questions to their teacher. Such questioning must not be permitted at all times during the lesson, or it will tend to break the continuity of the work. At certain stages, however, and particularly at the end of the lesson, it will be well to give the children every opportunity of satisfying their natural curiosity. Each question asked is, to the teacher, an encouraging proof of the interest taken in the lesson; and the more thoughtful ones give evidence as to the working of the minds of the children, and also serve, to an extent, as a measure of the value of the work done.

Of course it will frequently happen that even a young child will ask a question which the teacher cannot answer, but this is not necessarily a proof that the latter is not properly qualified for his work. Nature is so varied and so full of changes that even after many years of close and constant study of her productions and phenomena one is always finding some object which has not been seen before, or noting some phase which has never before presented itself; and it is always possible for a child to discover what a naturalist has never seen. But even so, a teacher should put himself in the best possible position to deal with the various questions the children may ask by keeping his knowledge as far as possible in advance of that which he desires his children to acquire.

Should it happen, as it sometimes will, that the teacher receives a question he cannot answer, he should not fear any loss of respect on that account. If the relation between the teacher and the class is such as should exist, the latter will never withdraw its confidence and respect because, occasionally, the former is unable to give an honest answer to a question asked.

It is not at all an uncommon thing to hear a teacher say: "One of my children asked me so and so, and I gave such and such a reply; was that right?" In a case like this the teacher is probably ashamed to admit that he does not know, and so he frames some kind of answer and presents it with a hope that it may possibly turn out to be correct. This should never be done. The teacher's information must be accurate, and he himself must be true.

Again, in order that a teacher may be able to carry out a nature course successfully, he must himself be a student of Nature. If he is to arouse enthusiasm in the children under his care, he must himself be an enthusiast. This same remark also applies, of course, to the other subjects he is called upon to teach; and thus we come to the logical conclusion that every teacher must be an enthusiast in everything he undertakes to teach. This is, as we know, almost impossible in the case of a teacher who has to deal with all the subjects belonging to a modern curriculum, but still there is no reason why the teacher should not do his best to make the nearest possible approach to this ideal condition.

In some schools an attempt is made to increase the quality of the teaching by allotting to each teacher a subject rather than a form or class. Thus each member of the staff is, or becomes, to a greater or lesser extent, a specialist in his particular work.

There is a great deal to be said in favor of this arrangement; for if, as should be the case, each teacher is occupied in dealing with his favorite subject, the energy and enthusiasm naturally put into the work must necessarily be greater.

This system, however, has at least one drawback. The teacher, having no fixed form of his own, but passing continually from one group of children to another, has not an opportunity of acquiring that intimate knowledge of the habits and dispositions of the children which is necessary in order to mold their characters.

Nature study seems to be one of those subjects for which a special teacher is more particularly advantageous; for while the majority of teachers possess a satisfactory knowledge of most of those subjects that form part of the ordinary curriculum of nearly all schools, the study of Nature has received but little attention until recently, and thus fewer teachers would consider themselves suitably qualified for dealing with it.

Reverting now to the subject from which we have slightly digressed, we next draw attention to the desirability of always encouraging the children to sketch what they observe, and thus to keep both eye and mind working together. Of course many of the attempts on the part of the children, and especially of the younger ones, to represent what they see will be very crude and inaccurate. That, however, is a matter of but little importance. It is sufficient that they have made a good attempt. The results will gradually, perhaps rapidly, improve as time goes on; and we may be sure that most children at least have observed the object before them much more closely than they would have done had they not been told to give a graphic representation of it.

Again, if the object selected for a nature lesson is to be thus represented by the class, the drawing need not necessarily be part of the lesson itself, but may form an entirely separate lesson in drawing, either on the same or another day. The two subjects, nature study and drawing, should run together; and it is of little importance whether the drawing lesson precedes or follows the corresponding nature lesson. If the former, the nature study will probably progress a little more rapidly because much of the observation has been previously done; if the latter, the drawing will be more accurate, especially in matters of detail, on account of the previous minute examination of the object during the nature lesson.

As regards the teacher's own drawing and illustrations we shall have many remarks to make; but we may set it down as a general rule that a nature lesson, based on specimens which have been distributed to the children for study, or on a large object placed before the class for the observation of all, requires but little blackboard illustration, if any at all. No sketch or picture should be presented that merely 'illustrates' that which may be observed in the object itself, not even if the former displays certain particular features more conspicuously than the latter. Let the children have the full opportunity of searching out the features for themselves. Do not attempt to save them any trouble, for this will deprive them of the pleasure of finding out for themselves. It is close observation that we desire to encourage, and, therefore, we do not tell them what they ought to see, but rather let them have the pleasure of telling, in their own simple language, what they have discovered.

At times, however, pictures are very useful aids. Thus, after it has been made clear that the general form of a certain tree must necessarily depend to a great extent on the arrangement of the buds as seen in the twigs placed before them, a picture of the whole tree may be shown as a means of demonstrating the conclusion; but even this is unnecessary and unadvisable where it is possible for the children to observe the tree itself within a reasonable distance from the school or their homes.

Diagrams are often useful to the teacher himself in assisting him to direct the observations of the children. It is often necessary to call special attention to some particular part of the specimen that is being examined, if only because it is advisable to secure some definite order in the work—to see that all the children are giving their attention to the same part at the same time. It is often somewhat difficult, especially with junior classes which are unacquainted with the names by which the parts of an object are denoted, to specify the particular portion requiring attention; but a diagram, even a very simple one, will enable the teacher to point it out immediately.

The same purpose may also be served by the use of a model instead of a diagram. Thus, in calling attention, in order, to the parts of a flower, a model of the flower, sufficiently large to be distinctly seen by all the class, will prove much more useful than the best of diagrams.

With the aid of such simple materials as plasticine, pieces of paper of various colors, wood splints, pieces of wire, etc., exceedingly useful models of various natural objects may be put together in a very short time.

Both diagrams and models should be used sparingly. They are not to be employed for the observation of the children, but as an aid to the teacher. They should be out of sight except at the short period or periods during which they are actually necessary, or the children's attention, which should be devoted entirely to the natural object before them, will be divided between the two, thus helping to destroy what should be the main aim of the nature lesson.

A thoughtful teacher can often foresee some of the difficulties that are likely to arise during the course of a lesson—difficulties that may require the aid of a blackboard sketch, and will prepare what is necessary beforehand; but even the most experienced teacher cannot foresee all that is required, and therefore he should be able to produce a satisfactory sketch, in the shortest possible time, to satisfy the exigency of the moment. Without such skill the lesson is liable to run slowly at times, and the labored production of a simple drawing will demand a pair of eyes that should be ever directed to the class and its working.

Really good pictures representing natural scenes and phenomena are very valuable both in connection with, and apart from, the nature lessons of the school, especially in populous towns, the children of which seldom have the inclination or opportunity of taking a ramble in the country. Such pictures enable the teacher to broaden the scope of his lessons, and to illustrate those casual, pleasant chats about the ever-changing drama of the seasons, and the general aspect of wood, wayside, meadow, moor, and mountain, that create a desire to stray from the crowded streets to open spaces where the realities of Nature may be enjoyed.

The instrument formerly known as the magic lantern and used for entertaining purposes, but now designated the optical lantern and recognized as a valuable aid to education, is an appliance to be found in almost all well-equipped schools. It is often employed in connection with the subject we are now considering, but its use is decidedly wrong if the pictures exhibited take the place of natural objects or illustrate scenes such as may be observed within a reasonable distance of the school.

However, the remarks made above concerning the use of good pictures apply, of course, to the use of suitable lantern slides. Beautiful photographs illustrating all kinds of natural scenes and phenomena are to be obtained in this form, and the use of the lantern has the distinct advantage that a number of pictures, magnified to suit the size of the school or class, can be exhibited in succession on the screen.

And here we must note the close relationship existing between nature study and geography, the latter being really a branch of the former, so that the rules laid down with regard to the illustration of nature lessons should be observed as closely as possible in the study of geography. Direct observation, carried on as far as may be in the open air, will certainly produce the most beneficial results on the minds of the children; and this may be supplemented by the use of good pictures, including photographs from Nature, exhibited either with or without the aid of the lantern.

The lantern may be made to serve yet another purpose in connection with nature lessons. It not infrequently happens that several diagrams are required for the purpose of aiding the teacher in his directions and explanations during a single lesson. In this case the necessary drawings may be made on small pieces of glass, instead of on the blackboard, and then projected on the screen as required. And it does not appear to be generally known that the classroom need not be darkened for this purpose.

If, instead of throwing the light on an opaque screen with the lantern at the back of the class, we have the lantern behind a translucent screen consisting of a sheet of tracing paper or tracing cloth, or even of a sheet of ordinary drawing paper that has been rendered translucent by painting it over with melted paraffin, and project the picture through it from behind, the light of the room need not be reduced any more than it is by letting down the ordinary window-blinds; and thus the teacher can make use of his diagrams while the children are still observing the natural object or objects placed before them.

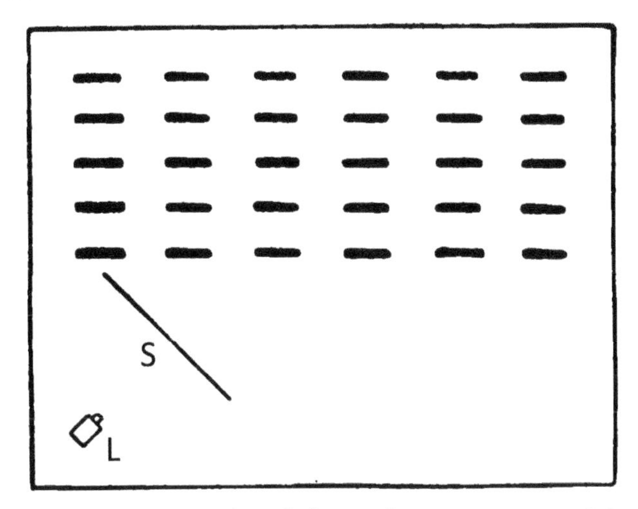

How to use the optical lantern without darkening the room. L: Lantern, S: Screen (Fig. 1).

With such an arrangement in a rather small classroom it will not be possible to throw a large disc on the screen; but then, in such a room a large disc is not at all necessary. The diagrams need not be any larger than the blackboard sketches for which they stand as substitutes, and thus a disc of about two feet in diameter will be ample.

What a stimulus, too, to the children, to encourage them to study and sketch natural objects at their own leisure, and then to allow them to project their drawings on the screen and to tell their mates of their discoveries and experiences! Give each child who desires it a little square of glass, with the few necessary instructions and, when the drawings have been brought in, note the delight with which the children exhibit their handiwork and explain what they saw, and the intense, stimulating interest displayed by the others as they observe what their classmates have discovered and accomplished.

An occasional half-hour spent in this way will do wonders in encouraging keen observation and in promoting accurate representation; and not only will the success of the experiment frequently come as a great surprise to the teacher himself, but he will sometimes find that the researches of his children include some little matters of structure or habit that he himself had not previously noticed, or give rise to some little thought or opinion which he himself can appreciate.

There are certainly a few little difficulties in the preparation of the simple lantern slides we have mentioned, and a few little knacks to be observed; but the latter will be quickly overcome if attention be paid to the hints in the chapter on Nature Lantern Slides.

Before quitting the subject of the uses of the lantern in nature study we would like to give one other illustration. Let us suppose that the children have received a more or less systematic training in nature observations as they passed through the junior classes, and that the course included, among other things, the study of various common plants and animals. As these children reach the higher forms they are in a position, from the knowledge gained, to arrange the various objects they have seen into natural groups—to plan out, with the aid of the teacher, an elementary system of classification. In such a case it would be well to recall the various observations made in the past by means of pictures thrown on the screen, thus aiding them in the useful exercise of classifying and grouping.

Returning again, for a moment, to the ordinary nature lesson of the school curriculum, we desire to say a few words concerning blackboard notes and recapitulations.

As regards the former, it should be definitely decided whether the notes are intended for the aid of the teacher himself, or for the observation of the children. If they are intended to be a guide to the teacher, and to consist of the headings and main points of the lesson, they are entirely out of place. In this case they should not be necessary, for the teacher should have so carefully planned his work as to require no such aid. And again, they distract the attention of the children from the object they should be closely examining, especially if they are written before the class and during the lesson.

If, on the other hand, the blackboard notes are intended for the sole use of the children, it is still difficult to see their value or of what they should consist. It may be said that all hard and unfamiliar words used should be written on the board. Not so. Hard and unfamiliar words should not find a place in a nature lesson. The descriptions and other statements are given by the children, in their own simple language, and technical terms should never be substituted by the teacher for the corresponding names or phrases of the child. Questions which turn upon words rather than upon things should always be carefully avoided. It does not matter much what a child calls any particular thing or part, providing the name given is fairly appropriate.

Our aim is to get the children practically acquainted with things, not names. Many a child has developed a great distaste for such a study as botany because the work set him was to learn the names of the parts of flowers and to learn to give descriptions in the technical expressions of certain textbooks. The effect would have been very different had he been taught to look upon flowers as living things with beautiful forms, lovely colors, and interesting habits.

It is usual to set apart a portion of the time allotted to a lesson for purposes of recapitulation, and this practice is often so rigidly observed that the notes demanded from young teachers are regarded as incomplete unless some provision for recapitulation has been arranged. This is quite unnecessary, and even inadvisable as far as our present subject is concerned.

Ordinary lessons of information require more or less repetition. The teacher's chief aim in such lessons is to impart to the children some of the knowledge he himself possesses, and a recapitulation serves to drive home the facts that have been given. But, as we have already pointed out, the purpose of nature study is not to give information on natural objects and phenomena, but to encourage careful observation and independent thought. Our aim is, or should be, to assist the children in making discoveries for themselves, and it is for this reason that we are careful to tell them nothing which they can be made to find out for themselves. Let the whole of the lesson be spent in these observations and discoveries, and you will find that the children do not readily forget what they have found out by their own efforts.

Should the teacher desire to ascertain how the minds of the children are working, he can do so by means of suitable questioning as the study proceeds; in fact, such questioning should form an important part of the lesson. And here we may note how closely nature study comes in touch with the teaching of English; for not only do the descriptions and thoughts of the children, expressed in their own words, form valuable exercises in oral composition, but nature study provides a wonderful wealth of material for the children's essays.

Nature reading books are sometimes used as a substitute for nature lessons. This is undoubtedly a very great mistake, for while there is no reason why descriptions of natural objects should not be read as much as descriptions of anything else for the general purposes of the reading lesson, it must be noted that the aims of the reading lesson are quite foreign to those of nature study.

Considered apart from the mere mechanical functions of the reading, the matter of the lesson simply tells the children what they may see, or what somebody has previously seen, while in the nature lesson they see for themselves; and the former explains those problems which, in the latter, are worked out in the minds of the children.

If a nature reading lesson is accompanied by the observation of the natural object which it describes, and if time is allowed both for the examination of this object and for questions and remarks on the part of teacher and children, it will still constitute a very feeble substitute for the real nature study lesson, for it will still possess the defects we have just mentioned.

Where nature study forms part of the school curriculum, however, it will be well to encourage the children to read, in their own time, any good books of travel, and the popular works of eminent naturalists; and such books may also be used to advantage in the ordinary reading lessons of the school. The teacher, too, may do much to increase the general interest in Nature by telling of his own experiences and discoveries, by imparting some of his book lore, and by giving occasional Nature stories and biographies, especially to the junior classes.

— 3 —

Outdoor Work

Up to the present we have been dealing more particularly with nature lessons as given in the school building, but the most valuable part of nature study is undoubtedly the outdoor work—the study of things in their natural surroundings; and advantage should be taken of every available opportunity of rambles in lane, field or wood; or, in the case of town schools that are too remote from wild Nature, in any neighboring parks and open spaces.

Such rambles may be frequently organized and personally controlled by the teacher, but it is by no means necessary that such should always be the case. In fact, the teacher should do all he can to encourage individual and independent observations, and should allow a little time occasionally in school for chats on the observations made by the children and for the asking and answering of questions relating to them.

Every school ramble must be arranged with some definite object in view, otherwise much valuable time may be lost in aimless wanderings and disconnected observations. Although we lay this down as a fixed rule, we do not, of course, wish it to be understood that objects of interest which lie outside the range of the proposed work are to be ignored. While we have determined on a particular series of observations, all related to some definite portion of our subject matter, we must be careful that we do not suppress the individual enthusiasm of the children; but, at the same time, we must be equally careful that the object of the ramble is properly carried out.

Thus, if we go out, on a certain day in spring, for the express purpose of studying the bursting of the buds and the folding and unfolding of the young leaves, we ramble from tree to tree in the course of our work; but as we pass from point to point in our journey, neither teacher nor child will close his eyes to the many interesting objects that thrust themselves on their view. During these intervals we note the early spring flowers—how and where they grow, observe the first butterfly of the season as it flies across our path, watch the queen humble-bee as she searches out a suitable spot for her nest after a long winter's sleep, and pause to look at the little lizard as it basks in the warm rays of the sun. So, at the end of the ramble we shall have carried out our object as regards the bursting buds, and also learned much concerning other interesting things.

Each child should be provided with a notebook and pencil for the purpose of recording what is seen. Even the youngest of the children should be encouraged to do this, and although the result may be very disappointing to the teacher, he must be satisfied, for the present, that an attempt has been made, and that every such attempt must have some effect in forming the habits of the child.

The nature of the entries will vary according to the age and capabilities of the class. Encourage all children to make a sketch of at least those objects which have been specially selected for the observation of the day, and see that all entries are made under their proper dates, so that they may be transferred, in the case of the senior classes, to a well-kept nature diary. The systematic entry of observations made is a matter of such importance in the training of the child, and is likely to be of such great value and interest in years to come, that we think it necessary to devote a short chapter exclusively to the consideration of the manner in which notebooks and diaries should be kept.

In addition to notebook and pencil, each child should be provided with a box in which to take home those objects that are required for a more detailed examination than could be given during the excursion, and any specimens that are to be preserved for future study. The children of senior classes will require a pocketknife. A few small trowels may also be necessary for the collection of roots that are required for the school garden; and a magnifying glass and a compass will be of great value in many cases.

We have just referred to the collecting of material during the progress of the nature study ramble, but it must be remembered that the object of the ramble is not the collection of specimens for the illustration of the lessons to be given in the school, but rather the close observation and study of natural objects in their natural surroundings; and we must be careful that the children do not develop into mere collectors, but rather into keen observers.

It is true, as we have already hinted, that it is often advisable to take home specimens for a more detailed study than could be given in the field, and we shall often meet with things of an imperishable nature that are of such interest that they may with advantage be given a permanent place in the reference museum of the school.

Furthermore, we shall find many living objects, both of the animal and the vegetable worlds, the growth and life histories of which are of great interest and provide favorable opportunities for a series of continuous observations and records. Thus, the roots of wildflowers may be transferred, in their earlier stages, to the school garden, in order that the future development of the plants, their flowers and their fruits, may be observed day by day; the fruits and seeds of trees and herbs may be secured for the same purpose; and the fronds of ferns may be collected for the spores, in order that the cultivation of the ferns and the observation of their interesting life histories may be closely and continuously observed in the classroom.

Also, to give a few similar illustrations on the animal side, a caterpillar, together with a sprig of its food plant, may be taken for the purpose of studying the interesting metamorphosis through which the creature passes to its perfect state as a butterfly or a moth; various small animals of a suitable nature may be secured for the observation of their habits in the school vivarium; and many species of aquatic animals may be collected for the school aquarium, in which they can, as a rule, be far more easily observed than in the natural pond.

If such material as the above is to be collected during the school ramble, the teacher should see that he or the children are previously provided with suitable accommodation for the specimens that are required. For plants and flowers a moderately large tin box, containing a little damp moss, will answer all purposes. For dry material, such as seeds, fruits of a non-succulent nature, fern fronds required for their spores only, and various objects of the mineral world, any kind of box, or even strong paper bags, will suffice; though it frequently happens that a special box with a loose packing of cotton-wool for delicate objects is extremely useful.

Small living animals are conveniently transmitted in wooden or tin boxes in which a few holes for air have been made with an awl; but if a tin box is to be used for the conveyance of active little animals, the holes should be made by pushing the awl outwards from within, so that there are no rough edges of metal projecting inwards to the injury of the occupants.

We do not recommend the preservation of animal and vegetable specimens for school nature study. Our object should be to create an interest in living Nature by the observation of living things and their ever-changing aspects, and for this purpose we do not require the aid of preserved specimens.

Dried plants and flowers are so unlike the original objects which they represent, that, although they may be of some use for certain scientific purposes, they should hardly be needed in a school for the young. Similarly, preserved animals are of no value in connection with our work. Nature is always interesting because she is ever-changing. Illustrate a nature lesson by means of preserved specimens and its interest is usually as dead as the illustrations.

We have often seen classes at work on nature study, both within the school building and outdoors, and have noted that, in almost every instance, the subjects studied belong almost exclusively to the vegetable world. This, we think, is a great mistake; for, in addition to marvels of structure and development common to both animal and vegetable beings, the former possesses the additional feature of the power of movement and the interesting habits resulting therefrom. And there seems to be no reason why the observations should be confined to living beings only.

The characteristics of the various rocks, and the soils derived from them, are worthy of some attention; also the movements and varying conditions of the atmosphere; and the face of the sky, with the movements, apparent and real, of the different heavenly bodies.

It is in the study of these latter objects and features that we realize the very close relationship between nature study and geography, the one merging into the other without any line of demarcation.

The study of plant life is hardly complete, even in its most elementary stages, without some appreciation of the relation existing between the character of the vegetation of the districts traversed and the nature of the soil. Many plants are so partial to one particular kind of soil that we may often note a marked change in the nature of the vegetation as, in our ramble, we pass from one kind of soil to another. As a single instance we may note how the beautiful foxglove, often so abundant on clay or gravelly soils, suddenly practically ceases to make its appearance as we stroll from these soils on to a chalky or limestone district.

Children have often to walk some little distance in order to reach the locality in which their studies are supposed to commence, and to wander still further while their observations are in progress. In such cases they should be taught to observe the nature of the ground covered—the various slopes and aspects, the principal features of the vegetation in different districts, and the characters of the soils on which they tread.

On passing a quarry or a railway-cutting they should be encouraged to observe the underlying rock from which the soil has been in part derived, and compare the former with the latter. They should note those localities in which the soil bears no relation to the rock beneath it, and thus be led to inquire into the origin and mode of formation of vegetable soils, and into the various ways in which certain soils are transported from one place to another by the action of water and other denuding agencies.

As soon as the children are sufficiently advanced, let them sketch a simple plan of the route taken, and enter on this, on both sides of the route, the general characters of the adjoining ground—the positions of hill, valley and stream, of field, wood and moorland. Encourage them also to mark, as accurately as possible, the spots where the principal objects of interest have been seen, and to enter any observed changes in the nature of the soil.

The elder scholars, after having become more or less expert in the preparation of rough plans as indicated above, may be taught the use of the pocket compass, and also simple methods of measuring approximately the ground traversed. Thus they become initiated into the art of map-making. Also, they should be encouraged to find their way about the neighboring country with the aid of a compass and the ordnance map of the district.

Very interesting observations may be made during a ramble along the banks of a river or small stream. The moist banks are the special habitats of certain water-loving wildflowers, shrubs, and trees; and in the stream itself we find several species of aquatic plants with a structure peculiarly adapted to their watery home.

In such a ramble attention should be called to the varying velocity of the stream at different points, and the relation which the velocity bears to the gradient of the bed and to the transverse sectional area of the stream should be worked out. Furthermore, the different kinds of material forming the bed of the stream should be observed—the stony character where the stream is rapid, the sandy bed where the current is not quite so swift, and the muddy bottom of the sluggish parts. Thus the processes by which the stream tends to reduce the level of higher ground, and to fill up the hollows, may be worked out. All the important features and functions of a mighty river may be observed, on a small scale, by the study of an insignificant rivulet.

Much valuable study may also be done, in the case of the schools of seaside towns and villages, during a ramble along the coast. Here we may observe the result of the denuding action of the sea, and watch the waves as they do their work; while the bare cliffs give us ample opportunities of studying the rocks of the district.

The sea cliffs, too, have their own special vegetation, as have also the salt marshes that are to be found on low parts of the coast. Some interesting flowering plants grow only near the sea, while others, that are common inland, become much altered in growth and habit when they find a home on the cliffs.

On the beach itself the children may observe some of the results of the mechanical action of the waves in the rounded outline of the lower rocks, the pebbles, and the particles of sand. Here, too, the movements of the tides should be noted; and the times of ebb and flow, as well as the limits of the advance and retreat of the water on different days, entered in notebooks for future reference. Side by side with these entries the condition of the moon at the same time should be noted. Thus the children are led to see that the hour of high tide and the amount of advance and retreat of the water are always the same for the same condition of the moon. So they are led to look upon the moon as the prime agent in the production of the tides, and are put in a better position to understand the theory of those movements when they are old enough to grasp it.

Marine life provides a wonderful store of material for nature study. Let the children observe the general features of seaweeds—their varying forms and colors, their mode of growth, the absence of flowers, roots and true leaves.

In the case of those larger species that are provided with air-bladders, let the children observe the plants as they hang over the rocks at low tide, and again their position when submerged. They will then be able to see the function of the bladders in supporting the plants in such a manner that they receive a maximum of light and a free supply of dissolved air, and thus they become acquainted with yet another example of adaptation of structure to habit and habitat.

At low tide they may examine the various forms of animal life—mollusks, crustaceans, worms, jellyfishes, sponges, etc, that live attached to the rocks; the various creatures—crabs, small fishes, etc., that conceal themselves beneath stones and weeds while they wait for the return of the water; and the many active animals that people the rock-pools.

Here they will have the opportunity of observing many examples of animals that are protected by hard external coverings; many also that are provided with ample means of defense and offense; and quite a number which are protected from their enemies, or enabled to lie concealed in wait for their prey, by a remarkable resemblance to their environment. Equally interesting and instructive are the varied organs of motion and locomotion possessed by the creatures of the sea—the fins of the fishes, the jointed legs of crabs, shrimps and prawns, the gliding 'foot' of the winkle and the whelk, and the swaying tentacles of marine worms and anemones.

Even on those less productive shores where there are but few rocks to afford attachment to weeds and give shelter to animal life, much may be gained by a careful examination of the line of debris washed up by the waves to form the high-water mark, and this is more particularly the case just after a storm. Many forms of both animal and vegetable life that live and grow only in places that are perpetually covered with water are detached and thrown on the beach by the waves; and thus much may be learned of marine life during a stroll along the line of material which marks the limit of the flood of the recent tides.

If possible, a little time should be set apart occasionally, say about once or twice a week, for a general chat on the observations of both teacher and children, made during the last few days. This will greatly encourage the children to observe, and probably add much pleasure to their work; while, at the same time, the relating of their experiences will give the teacher very favorable opportunities of developing their power of expressing themselves in correct English.

Again, the teacher will often be able to make use of the children's observations out of school hours as a basis for definite nature lessons in the school building. Suppose, for example, it is proposed to give a lesson on some domestic animal that could not be conveniently studied within the schoolroom. Then, tell the class, a few days previously, to watch the particular animal closely, whether it be in the home, the field, the stable, or in harness, and to be prepared to give a description of its structure and habits. It would be well in such an instance, for the teacher to give some definite instructions as to the principal observations that should be made, e.g.:

1. The general form or build.
2. The character of the natural covering.
3. The limbs, especially in motion.
    a. The movable joints.
    b. The feet and their hoofs or claws.
    c. How far the limbs resemble, and how far they differ from, our own.
    d. How the animal moves about.
4. The head and neck.
5. The ears (compare with the human ear).
6. The eyes: where situated; lids and lashes.
7. The mouth.
    a. The lips.
    b. If possible, the teeth.
    c. How the animal feeds. Its food.
8. In all matters enumerated above, how the animal is peculiarly adapted to its habits and mode of life.

Then, when the time appointed for the lesson has arrived, the teacher will receive from the class all the observations made and the conclusions at which the children have arrived. He will not give information himself, as a rule, but rather encourage the children to observe again in matters where their observations have been imperfectly made. Nor will he offer explanations too freely, but cause the children, with as little aid as possible, to work out for themselves the little problems concerning the relation between structure and habit.

Such a lesson ought to be quite as valuable as one in which the object selected is examined in the presence of the teacher himself.

We may cite an example of another object that may be dealt with in much the same way—the obnoxious but interesting little housefly. This insect is much more conveniently observed at home than in the school. It should not be captured, but rather observed at liberty. Encourage the children to note:

1. The divisions of its body—the division of the body into distinct segments.
2. The large eyes, little feelers (antenna), and the sucking organ belonging to the first segment or head; also the food required by the fly and the manner in which it feeds.
3. The legs: where situated; how the fly walks, and its power of walking on very smooth surfaces, even in an inverted position.
4. The number and nature of the wings, and the wonderful power of flight.

In this instance the aid of the teacher's knowledge will be necessary in explaining exactly how the fly feeds, and why it is unable to devour food in the solid state; also, how it is enabled to walk on perfectly smooth surfaces. Diagrams or better photographs from Nature might be shown to demonstrate the wonderful structure of the foot and the proboscis. The subject might also be extended by rearing some flies (the common meat-fly or blow-fly is, perhaps, best for the purpose) in a suitable cage (see page 217), in order that the children may be able to trace the whole life history and metamorphosis.

There are many other nature subjects that lend themselves particularly well to this mode of treatment, and they have the distinct advantage that they greatly assist the teacher, in that they give the children useful and interesting employment for their leisure hours, and thus allow of more actual study than would be the case if all observations were made under direct supervision.

In this chapter we have endeavored to point out a little of the work that may be done outdoors, but much more will suggest itself to a teacher who is interested in his class, the nature of the work varying according to the situation of the school.

Where nature study has formed part of the school curriculum, and where the subject has been taken on the lines we have laid out, there will be but little fear that the children will cease to interest themselves in their surroundings as they get older and eventually leave school. Often we find them organizing themselves into little societies or clubs for the express purpose of continuing the work that has given them so much delight in the past, and it will be well if the teacher does his best to foster this tendency by helping the elder scholars and the old boys and girls to establish, organize and maintain a field club or natural history society in which they can continue the study commenced in their younger days.

It must not be supposed that this mode of treatment is exclusively adapted to the study of animal life, for it is equally applicable to all branches of Nature. The general study of our forest trees and shrubs, the habitats of flowers, the general aspect of hedgerow, field and wood, the study of atmospheric phenomena, the changes of the moon, and the movements of the various heavenly bodies are all suitable employments for children under the guiding hand, though not necessarily under the direct supervision, of the teacher.

In dealing with the above and other subjects it is hoped that the teacher will derive many useful hints and much practical aid from our future chapters treating with the school aquaria, vivaria, terraria, garden, and various other appliances for the continuous observation of living and other things either in the school or in the playground.

— 4 —

Spring Studies

A. General Remarks

Spring is the season of the re-awakening of Nature, and is therefore the best time for the commencement of a nature study course in schools.

At the beginning of this season the 'stocks' of the hedgerow plants, that have remained dormant throughout the winter, produce their new leaves, of a fresh green color; and these, together with the tender seedlings that have grown from the self-sown seeds of the previous summer and autumn, push their way through the old withered herbage, rapidly hiding the latter from view. Similar changes are taking place in pastures and waste places, and the spring foliage is soon relieved by the appearance of the early flowers, most of which develop rapidly from stores of food that were laid up by their plants before the winter set in.

In the woods the winter buds begin to burst, some of them exposing the young foliage leaves of the coming summer, and others giving rise to flowers that either appear before the leaves of the same tree, or expand in company with them; while beneath the trees we see the early spring flowers that must necessarily bloom before the foliage above is sufficiently dense to shut out the sun, and the multitudes of baby forest trees shooting through the leafy soil in the neighborhood of their parents.

The animals that have been dormant throughout the long winter are now called forth to new life by the warmth of the spring sun; and others, which have spent the greater portion of the cold season in sheltered holes and corners, now resume a life of activity. Birds are busily engaged in building their nests and tending to the wants of their nestlings, and the feathered friends that left us for warmer climes at the fall of the year now return and make preparations for the accommodation of their offspring.

In the fields the lambs are skipping; the ponds and streams are being rapidly restocked with new life; and the rock-pools between the tide-marks on the seashore are being re-peopled with a variety of curious animal forms. In the country, and even in town to a certain extent, human nature responds to this reawakening of life and activity, and various occupations that were retarded by the winter's frosts are now taken up with renewed vigor.

In the present chapter we shall call attention to the principal objects and events of the spring season, dealing with different departments of Nature in turn, and endeavoring to introduce more particularly those subjects which are suitable for a school nature study scheme.

B. Plant Life

I. Winter condition of trees and shrubs

Bole of the Oak (Fig. 2)

During late winter and early spring we should make a special study of the winter condition of the common forest trees and hedgerow shrubs, while they are still in their dormant winter condition.

As long as the air and the soil are still cold, the roots are inactive. In fact, cell-activity is practically suspended in all parts of the tree or shrub, and the circulation of the sap and all the functions that result therefrom are interrupted.

The trees being bare, we have now our most favorable opportunities of studying the various modes of branching—the peculiarities in the disposition of the branches and twigs by which we are able to recognize the different species in the absence of their leaves and flowers. We stand at such a distance from each tree that we can readily command the whole, and endeavor to fix in our mind the general features above referred to; and, passing from one species to another, we note carefully any points of resemblance as well as of contrast.

Bole of the Birch (Fig. 3)

Those who are not well acquainted with trees and shrubs in all their different phases will often fail to fix the identity of certain trees in their winter condition, even though the same species can be recognized with ease when in flower or in leaf. In such cases considerable help may be obtained by reference to photographs or other accurate representations of the trees in question; but, failing this, those species which present a difficulty should be watched at intervals as the season advances. This latter method is undoubtedly far preferable to the use of pictures, for it enables us to trace the trees through those interesting stages which mark the return to active life, while our observations are rewarded by the pleasure and satisfaction that we must all feel when a revelation is made—when the thing previously unknown reveals itself as an old and familiar friend.

After observing a tree at a distance, approach it with the object of learning the nature of its bark and the character of its buds. The former consists of a protecting covering of cork, composed of dead cells—cells which originally contained sap, but now only air. A new layer of cork is formed each year, and this is sometimes added to the accumulation of outer bark produced in previous years, so that, in an old tree, the protective covering is very thick. As a rule, too, a thick bark is very rugged, being divided into patches that are separated by deep, irregular furrows, as in the case of the oak and the elm.

This is due to the fact that the bark, being dead, has no power to produce new growth to accommodate itself to the increasing trunk and branches, and is therefore fractured by the outward thrust. Some trees, on the other hand, like the birch and the plane, are almost continually shedding the older and outer layer, which peels off while a new layer is being formed; and thus the bark of these trees is seldom very thick and rugged.

The study of the so-called 'winter buds,' which are really formed during the preceding summer, is both interesting and important. Not only may all our forest trees be readily recognized by the form and color of the buds, but it should be observed how the arrangement and development of these buds determine the general form and mode of branching of the tree.

A bud, it must be remembered, is a young branch. In several trees, including the beech, poplar, willow, and lime, there is a single terminal bud at the tip of each twig, and also lateral buds along the twig, arranged either alternately or in pairs. Imagine these buds to have developed into branches, and these branches in turn to have produced similarly-arranged buds and branches, and so on for several seasons in succession, and you have a picture of the tree as far as the general arrangement of its branches is concerned, but this picture may be more or less modified by the destruction or imperfect development of some of the buds, and by the direction in which the twigs grow.

The single terminal bud at the tip of each twig continues, by its growth, the general direction of that twig; and each lateral bud produces a twig resembling the branch from which it grew. Compare the above condition of things with that which obtains in the case of the oak. Here, at the tip of each twig, there is a cluster of from two to five or six buds, one or more of which are often much more strongly developed than the others. If only one of these buds develops in the spring, the twig formed is seldom in a straight line with the one that gave rise to it; and if two or more grow, they produce new twigs diverging from one another. Thus, by studying the distribution of the buds on the oak tree, we are enabled to explain the crooked and gnarled appearance of the old kings of the forest.

We may take yet another example—that of the sycamore. On the twigs of this tree we note that the terminal buds are frequently in pairs, while the lateral ones are also in opposite pairs. With such an arrangement of terminal buds it is, of course, impossible for the new twigs to continue the direction of the older ones; for if both develop a fork is formed. Also, the two lateral buds of each pair will give rise to two branches diverging from the parent twig. Here then, we have an explanation of the rather crooked branches of the sycamore, the frequent forking, and the common threefold divergence of the same. The ash also has its lateral buds opposite, but each twig has usually a single terminal bud.

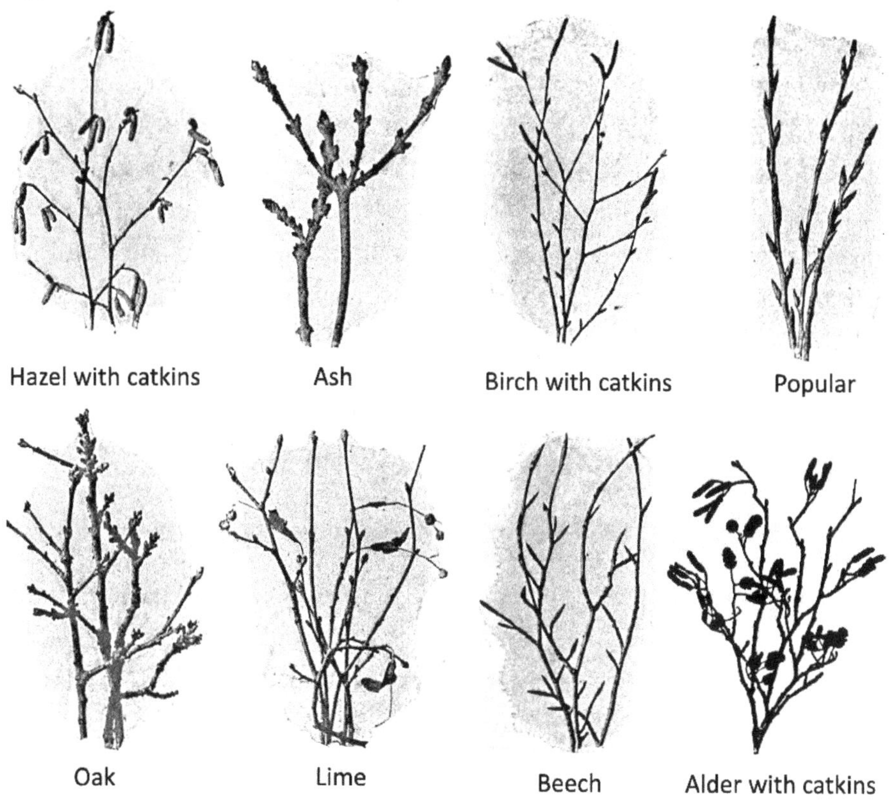

Trees in winter or early spring (Figs. 4 & 5)

Children should be taught to observe the above and other characteristic features of our common forest trees in winter, and the teacher should encourage and help them to think out the various problems connected with their growth—how it is that some trees have a thin, smooth bark while others are provided with a thick, rugged covering; and how the tree came to be of its present form as regards the arrangement of its branches.

II. The bursting of the buds

The best time to study the buds of trees is the period immediately preceding the commencement of the expansion. This period will, of course, vary considerably as regards the different species, so that the whole study will extend throughout, perhaps, the whole of March and April.

Let the children observe the times of the appearance of the first leaves of different trees, and enter these times in their nature diaries. Continuing this, year after year, they will see not only that trees of the same species will be earlier or later, in the same season, according to the situation of the individual trees and the soil in which they grow, but also that the season itself varies in different years, sometimes calling forth the young leaves very early, and sometimes retarding their appearance to a much later date. They will also learn that some species are generally earlier than others, and that the order of the bursting of the bud, as regards different species, is not always the same.

All these variations are both interesting and instructive, especially as they lead to the consideration of temperature and atmospheric conditions, and help the children to better understand the circumstances most favorable to vegetable growth.

Buds should be examined in their winter or dormant condition previous to the observation of the opening and the expansion of the leaves; and particular attention should be called to the impervious character of the scale-leaves that surround and completely enclose the delicate structures within, and which prevent loss of moisture by evaporation at a time when the tree is deriving but little water from the soil. Some large buds, such as those of the horse-chestnut, may be cut longitudinally with a sharp knife in order to show the parts of the future branch—stem, leaves and flowers—in situ; and transverse sections are equally instructive, and are especially useful in showing the manner in which the young leaves are folded within the bud.

The examination of such sections, as well as the observations of the opening of the buds, will show that some are destined to produce branches that bear foliage leaves only; that others are to give rise only to floral branches; and that some are to develop into branches bearing both foliage and floral leaves.

If the observations of opening buds are made only outdoors it is possible that some very interesting stages may be missed on account of breaks that occur through unfavorable weather and other circumstances; but a good many buds of forest trees and shrubs open very readily if the twigs bearing them are placed in a vessel of water in the schoolroom, and these give very favorable opportunities for a continued series of observations and for the drawing of series of sketches.

Buds of the opening Horse Chestnut (Fig. 6)

We have already referred to the dissection of buds with the object of studying the manner in which the leaves are folded within; but this folding can be seen remarkably well in most buds just as they begin to expose the leaves, and may, therefore, be studied from the specimens kept in water. Yet it will be wise to extend these observations outdoors when this can be conveniently done, not only because we may thus have the opportunity of seeing a greater variety, but also because the expansion of the buds may be observed to later stages and under more natural conditions than would be the case with indoor specimens.

Particular attention should be called to the folding of the young leaves as they make their first appearance, the method of folding being, of course, the same as that which obtained while the leaves were entirely enclosed within the bud, and revealed at this earlier stage by transverse sections only. It must be observed, too, that most leaves retain their folds for a period after they have freed themselves from the cover of the protecting scale-leaves that formerly enclosed them, this being a matter of considerable importance; since the epidermis of the tender leaves is as yet very thin, and readily permeable to water, the sap within would rapidly pass out and evaporate, especially when the air is dry, thus causing the young leaves to shrivel and die. The retention of the folds, however, prevents a free exposure to air-currents, and consequently retards loss by evaporation.

Again, it will be observed that many young leaves are further protected from the drying influence of air currents by a covering of downy or silky hairs. Note, for example, the young leaves of the horse-chestnut and the beech. The leaf of this latter tree is characterized by strongly marked parallel veins branching off from the midrib on either side; and, when very young, it is folded in such a manner that the only parts exposed are these veins and the margin. But there is a line of silky hairs on each vein and on the margin, so that the folded leaf is completely surrounded by a hairy covering.

As the leaf grows and expands, the lines of hairs are separated and the leaf becomes more exposed; but the epidermis is now less permeable to moisture, and the former protection is no longer necessary; and, the function of the hairs having been duly performed, they are gradually shed, so that the old leaf of late summer bears hardly a trace of them.

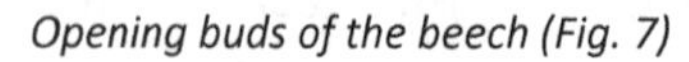

Opening buds of the beech (Fig. 7)

Young leaves of the horse chestnut (Fig. 8)

Evaporation of moisture is also greatly accelerated by full exposure to the rays of the sun, but the retention of the folds in young leaves prevents such exposure. In this connection it is also interesting to note how some young leaves, quite apart from their folding, dispose themselves in such a manner that they are more or less protected from the direct rays of the sun. The leaves of the horse-chestnut are at first densely folded, and covered with a mass of woolly hair; but, when they expand, they are still in danger of suffering a serious loss of moisture on a sunny day, or even on a windy day when there is no sun; and so they hang vertically from the tips of the branches, thus affording themselves and one another a certain amount of protection from both sun and breeze.

As a rule, those parts of a plant that have only a temporary function to perform are shed as soon as their work has been completed. This is certainly the case with the scale-leaves which protect the buds during the winter and early spring, and with the stipules that are required for the protection of leaves only during their early stages.

Children should be taught to note such phenomena and, on seeing, for example, the shower of deciduous stipules falling from lime trees and covering the ground beneath during April, to inquire, What is this falling? Where does it come from? Where did it grow? What was its use? Why is it now shed?

Twig of the Lime in spring (Fig. 9)

The trees and hedgerows in spring afford abundant opportunities of studying an enormous variety of opening buds and the interesting phenomena associated with their expansion, but it is hoped that the few instances quoted will be sufficient to show the teacher something of the nature of the work that may be accomplished.

During the winter months, long before even the earliest of the buds begin to burst outdoors, very interesting and instructive observations may be made on growing bulbs, such as those of daffodils, bluebells, tulips, onions, etc.; and afterwards, or even at the same time, while the structure of the buds of various trees and plants is being studied, comparisons should be made with the object of showing that a bulb is really a bud. Longitudinal sections of bulbs should be exhibited side by side with similar sections of the buds of a tree, such as the horse-chestnut; and it should be shown that the former are buds with thick, fleshy leaves, all attached to a mass of hard substance beneath which is really a shortened stem.

These bulbs should be grown under varying circumstances in order that the children may become acquainted with the conditions necessary for healthy development. The following notes give an outline of some of the more instructive experiments:

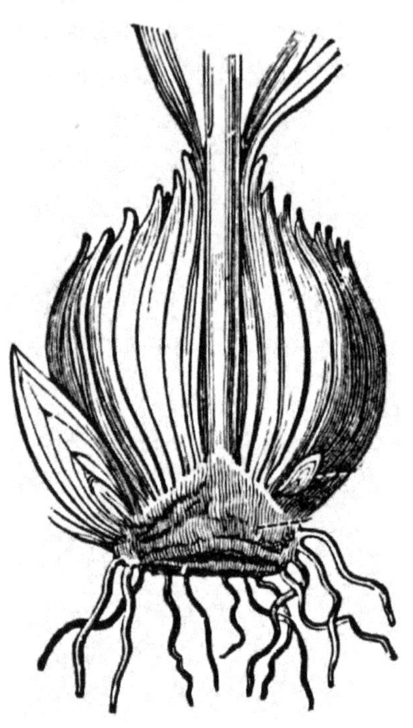

Section of an Onion (Fig. 10)

1. Bulb grown in a good soil and under favorable conditions as regards warmth, light and moisture. The plant commences to grow at the expense of food material stored in the thick, fleshy leaves. Roots are soon formed, and these absorb water and dissolved mineral food from the soil. The leaves appear, absorb carbonic acid gas from the atmosphere, and manufacture new material from the elements of the food obtained from both air and soil, thus assisting the growth of the plant. The plant then reaches its flowering stage, and the flowers give rise to fruit and seed. After the flowering and fruiting stages are over, the leaves continue to manufacture mineral food, which is stored in the bulb, or used in the formation of new bulbs, thus providing for a new plant or plants in the following season.
2. Bulb in good soil, with the same amount of heat and light, but no moisture.
3. Bulb grown in the same soil, with sufficient moisture and warmth, but kept in the dark. The unhealthy, straggling, white plant produced will clearly demonstrate the necessity of light for this kind of vegetable growth.

4. Bulb grown in the same soil, with light and moisture, but kept in a cool, exposed situation. The result—a much slower growth—will show how heat affects the development of the plant.
5. Bulb grown in water only (or in wet sand, sawdust, or fiber). The plant flourishes to the flowering stage, but not sufficient food is obtained to produce strong bulbs for the following season.
6. Bulb grown in water (or wet sand, sawdust, or fiber), and supplied with a proper amount of mineral food or 'fertilizer.' A strong plant produced, and sufficient food stored up in the bulb (or new bulbs) for new plants in the following spring—the result as good as when the same kind of bulb is grown in a good soil.

Other experiments, equally valuable as a means of education, will suggest themselves to the teacher. It will be well to let the children themselves plant the bulbs for the intended experiments, and to attend to them throughout. Children take a greater interest in what they do themselves than in what is done for them. All the class should also take a part in making and keeping the records of the results; and each child should be called upon to give his opinions and conclusions as regards the results obtained.

III. The movements of the sap in plants and trees

Various simple experiments should be performed by the children with the object of learning how water is absorbed by roots and transpired by the leaves; and also how the sap flows in a living plant. For full descriptions of such experiments the reader should refer to suitable botanical works, for we can do no more here than briefly refer to them.

1. Remove a plant from the soil without injury to the root. Wash the soil from the root, stand the plant in a graduated jar or bottle of water so that the root is quite submerged, and then close the neck of the jar with some impermeable material (such as a good cork cut longitudinally and grooved to grasp the stem) to prevent evaporation. Note the rate at which the water disappears from the bottle, and how this rate varies with different atmospheric conditions and with the amount of light.

2. Cover this plant with a bell-jar, and observe the condensation of water that has been transpired as vapor from the leaves.
3. Cut away a complete ring of the outer bark from the branch of a tree or bush. The branch still thrives, thus proving that the function of the outer bark is protection only.
4. Cut a deeper ring around a second branch, dividing both the outer and the inner bark (bast), but not penetrating to the wood, and note that while the branch still thrives and grows thicker above the ring, there is no increase in thickness below it.
5. In the case of a third branch, preferably an old one, cut a ring through both outer and inner bark, and also through the outer rings of wood. This branch dies.

From experiments 4 and 5 the children will (perhaps aided by the teacher) come to the conclusion that the sap rises from the roots through the outer or sap wood, that new material is manufactured in the leaves, and that this new material is carried downward through the vessels of the inner bark to add to the growth of the tree.

Other suitable experiments may suggest themselves to the teacher, and in all cases the children should, with as little assistance as possible, work out their own conclusions from the results.

IV. Observations and experiments connected with food storage

Many plants prepare and store considerable quantities of food material for use at a future period. Some of these are biennials—plants which do not, as a rule, produce their flowers and seeds during their first season, but lay by a quantity of food that enables them to reach maturity early in the following year. As familiar examples we may mention the turnip, parsley, parsnip, carrot, wild lettuce, and the spear thistle, all of which store food in their large, fleshy roots. Other plants which store food are perennials, and these usually lay up their supplies for future use in thick, fleshy roots, or in thick, underground stems.

The plants referred to above may be employed for various interesting experiments. Thus, if the tops of carrots, turnips, parsnips, etc., be cut off horizontally and placed in saucers of water, they will produce beautiful tufts of leaves; and potatoes (modified underground stems) that have been stored through the winter will develop stems and leaves from their buds (the 'eyes') in the spring. In all these instances a certain amount of growth will take place even without the aid of added water, since the parts contain large supplies of moisture in addition to solid food; but the development may be carried much further if water is used in the experiments.

Some of the above may be grown in a dark box or cupboard, in order to demonstrate the value of light to a growing plant. The comparison of a potato grown in the dark with another that has been exposed to a good light is particularly interesting and instructive.

V. Study of the germination of seeds

The study of the germination of seeds should be preceded, in early spring, by the examination of the seeds themselves. Procure several suitable kinds of seeds, such as beans, peas, mustard seeds, maize, and oats, soak them in water for about twenty-four hours, and then distribute them among the children for examination.

After the exterior has been observed, cause the skin to be removed, and direct attention to the young root, the bud, the seed leaves or cotyledons, and to special food, stores extraneous to these parts, if such exist. It should be made clear that the seed is really an embryo plant, awaiting favorable conditions for its development.

Having observed the general structure of the seeds, a moderate number should be sown under varying conditions, the growth of the young plants carefully watched, and records kept. The children should take the active part in all this. They will sow the seeds, tend the young plants, keep written records of the growth, and make dated sketches of the natural size, under the guidance of the teacher.

It will be necessary to have a number of seedlings that may be removed from their bed at intervals for the examination of the roots. These may be grown in wet sawdust, from which they can be readily removed without injury.

In cases where the earlier stages only are required, plain water only is necessary; but if the plants are to reach a mature stage, the seeds must be grown in a good soil, or watered with a nutrient fluid such as that given below.

It is important that the children become practically acquainted with the conditions essential to the healthy development of a plant, and to this end it is necessary to give the seeds and seedlings various modes of treatment. In this connection the following hints may be of some service:

1. Count the seeds sown in some instances, in order to ascertain the proportion that grow.
2. Sow some in sawdust, sand, or soil, and keep them quite dry, to demonstrate the necessity of water.
3. Sow others in sawdust or sand, and keep them moist with pure water only.
4. Place seeds of the same kind in a similar bed, but water them with a nutritive solution made as follows:

    Common salt .... 1 part
    Potassium nitrate .... 2 parts
    Calcium sulfate .... 1 part
    Calcium phosphate .... 1 part
    Epsom salts .... 1 part

    Dissolve about a quarter of an ounce of the above mixture in a gallon of water, and add one or two drops of a weak solution of iron perchloride.
5. Sow some in a good soil, and give them the normal treatment for plants in general.
6. Sow seeds in a good soil, or in sand or sawdust kept moist with the nutritive solution, but keep them always in a dark box or cupboard.

Perform the above experiments with seeds of the same species, and let the children enter the date in their nature diaries, and make complete records of the results as recommended above.

In the case of experiments 4 and 5, if the seeds selected are those of a forest tree or of some other perennial plant, the growth and the observations thereof may be continued season after season for years in succession; and very great interest will be aroused by the re-awakening, each spring, of the tree that was dormant and apparently dead during the preceding winter. If you would have your children interested in the nature and growth of forest trees, probably nothing will secure that end better than the formation of a little nursery in which trees are reared from seeds that have been gathered, planted and tended by the children themselves.

The results of experiment 6 should be very carefully compared with those of 4 or 5, the seeds being of the same species, and treated in exactly the same manner except as regards light. Here the children should not only make comparisons of the general characters of the seedlings produced, but also make measurements of the plants at regular intervals, and compare the average growth of the plants in one set with that of the plants in the other.

If suitable balances and weights are available, they may also take the average weight of the seedlings in the two sets at corresponding intervals, with the object of noting the amount of solid matter built up; in which case the plants must, of course, be thoroughly dried in a slow oven previous to the weighing.

We have already referred to the seed as a plant in embryo, containing the parts of the future plant distinctly visible though, as yet, imperfectly formed. In this respect the seed of a flowering plant differs from the spores which give rise to ferns, mosses and other flowerless vegetation. These latter are very minute cells that contain not the slightest trace of the structure of the plant which produced them; but the plants are often easily reared from the cells, and their life histories are particularly interesting. A method of propagating ferns from the spores will be found in the next chapter.

VI. Spring wildflowers

The study of the spring wildflowers will, of course, form a most interesting part of the work of this season, but we must be careful here not to adopt the practice, far too common, of confining our attention to observations of the form and color of cut or gathered flowers.

The Wood Anemone (Fig. 11)

Flowers are certainly very beautiful, and present a marvelous variety of form and color, and we should by all means make use of them in the cultivation of an aesthetic appreciation of all that is beautiful; but we need not trouble about those details of structure by means of which the systematic botanist distinguishes between genera and species. A flower may have four sepals or five, its petals may be united or distinct, its stamens below the pistil or above it, and the pistil superior or inferior.

The Lesser Celandine (Fig. 12)

The observation of these features will arouse but little interest, and, if the observance of them is accompanied by the use of the technical terms with which botanists are so familiar, the effect will probably be to make the flower-study a distasteful drudgery. Rather deal with plants as living things with wonderful habits and marvelous life histories.

Let our questions be: Where does it live? How does it grow? Why does it climb? How does it climb? How is it protected? What is the advantage of this or that particular form, color, or habit? etc. It is far better to encourage the continuous, thoughtful observation of the commonest weed in a neglected corner of a garden or on a wayside bank, than to make a study of the most conspicuous flower apart from its accompanying growth and its natural surroundings.

It follows, from what has been said, that the study of flowers should be conducted, as far as possible, outdoors. The notebook should now be in constant use for the purpose of recording the habits and habitats of the plants observed. If a nature diary has not been previously in use, the early spring is the best time of the year in which to encourage the children to start one.

Although we do not recommend any attempt at the classification of flowers on the part of young children, all striking resemblances between different species should be observed and noted; and the observations made will result in a store of knowledge which, in future years, may form the basis of a very elementary classification, confined exclusively to the plants that have been examined. Thus, the common buttercups may be compared with the water crowfoot and the lesser celandine; the stem, leaves and flowers of the dead nettles with the corresponding parts of the bugle and the ground ivy; the cuckoo flower with the wallflower and the garlic mustard; and the bluebell with the daffodil.

Colt's Foot (Fig. 13)

White Dead-Nettle (Fig. 14)

During early spring we meet with a large number of flowering plants that are well in leaf, but which, as yet, show no trace of the future flowers. Many of these are probably unknown in their present condition, even though they are easily recognized when in flower. Encourage the children to note the soil and situation in which they grow, and then to take the plants home and set them in a garden, preserving, as well as

The Ground Ivy (Fig. 15)

possible, the conditions natural to them, so that their progress and development may be watched up to the fruiting stage.

At first, while the weather is still rather cold, but few flowers are to be seen, but the few that appear are all the more conspicuous and interesting on that account, and will therefore receive their full share of attention.

The Elm in Flower (Fig. 17)

The Dog's Mercury (Fig. 16)

Some of the very early spring flowers are hardy species that bloom practically all the year-round, such as the shepherd's purse, chickweed, groundsel, and the red dead-nettle; and these are followed by, or accompanied by, the furze or gorse, the hazel, and the pretty little barren strawberry which is so often confused with the edible wild strawberry. But all the above flowers may be regarded as winter, rather than as spring, blossoms, inasmuch as they commence to bloom before the former season is at an end; and for this reason we give illustrations of a few of them, for purposes of recognition, in Chapter 7, dealing with winter studies.

With the advent of spring we renew our acquaintance with the wood anemone, lesser celandine, violet, colt's-foot, primrose, daffodil, the yew blossom, and the annual meadow grass. In the case of the colt's-foot special attention should be drawn to the fact that the flowers appear before the leaves; and the exact position of the clusters of flower-heads should be noted so that, in the following month, a second visit may be made to the spot in order to see the globular, hairy 'clocks,' and also the leaves, which are now commencing to cover a large patch of ground.

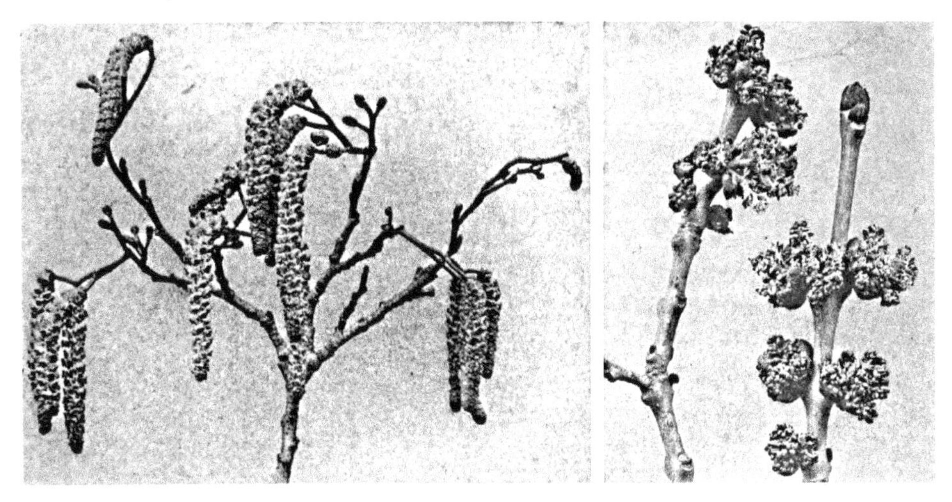

The Alder in Flower (Fig. 18) *The Ash in Flower (Fig. 19)*

During April quite a large number of common flowers will make their appearance, and the notebook and diary will be brought into constant use. The wayside is now brightened by the garlic mustard or jack-by-the-hedge, dove's-foot, crane's-bill, dandelion, white dead-nettle, ground ivy, and the dog's mercury. In meadows we find the common daisy, cowslip, the pretty little field woodrush and the fox-tail grass; and, on banks, the early forget-me-not.

This same month is still further interesting as being the period during which a number of our common trees and shrubs are in bloom. The willows, poplars, and alder are rendered conspicuous by the appearance of their catkins while the leaves are still hidden within their buds. The tops of the elms have a somewhat fluffy appearance, due to their clusters of small flowers, which are often absent on the lower branches. The flowers of the ash form dark purple clusters on the yet leafless branches. The oak, birch and hornbeam display their drooping catkins among the developing leaves which appear at the same time. The leafless sloe or blackthorn is thickly covered with its pretty, white flowers; and the yew still bears its little blossoms beneath its dark, narrow leaves.

The Bugle (Fig. 20)

From April onward the succession of new wildflowers is so rapid that we forbear even to mention their names. The teacher who has not spent some years in the study of the wildflowers and trees will, of course, frequently meet with those which he cannot name; and during his school rambles he will often find some, even common species, which he has not noticed previously. This need not discourage him, for the object of school nature study is to become acquainted with things rather than with names, and it is possible for one to know much of the interesting habits of flowers without even knowing what these flowers are called. It is a pleasure, however, to be acquainted with the names of the natural objects we meet with, and these may be gradually acquired by frequent reference to good books in which the descriptions and representations of the flowers are given.

As the various flowering plants and trees progress, they should be observed till they reach their fruiting stages, in order that the nature of the fruits and the modes of dispersion of the seeds may be studied. But few plants reach this stage before the season is well advanced, and consequently this portion of the work will be continued into the summer and autumn, and will be considered in its proper place. The photographs of some of the commonest spring flowers interspersed in this section will serve for the identification of species that are not already familiar to the reader.

C. Animal Life

I. General remarks

Some animals spend the whole of the winter and early spring in a perfectly dormant condition, and many others are more or less inactive during this cold period, venturing out of their hiding places only when the weather is mild.

In many cases the hibernating creatures may be readily found in their winter quarters, and these should be searched out with the object of learning what they do to protect themselves from the inclement weather. Thus, snails may be found in holes and snug corners almost everywhere, generally a number together, as if to derive some benefit from their companionship, each one with the aperture of its shell closed by a thin membrane formed from the slimy secretion (mucus) which exudes from its soft skin and from its mouth.

Also, beneath the surface of the soil, under fallen leaves, in decaying wood, and in various other suitable hiding places, we may meet with hibernating spiders, centipedes, woodlice, and many species of insects, either in the immature or mature state, awaiting the invigorating warmth of the spring sun.

The Marsh Marigold (Fig. 21)

Many persons have a great antipathy to a number of creeping and crawling animals, including the lower vertebrates (reptiles and amphibians) and the majority of the invertebrates; nevertheless these same creatures are well worthy of study, for they all have interesting habits, and many of them are really beautiful objects. Those who find them repulsive lose much, and the teacher should strive to overcome any such dislike he may have towards them, and to destroy any such aversion on the part of the children.

The mere suggestion of a lesson on the snail, or on the earthworm, is alone sufficient to create a feeling of horror on the part of some children, but if such a lesson succeeds only in removing this feeling no mean object has been attained. A good lesson on such an animal will seldom fail to have this effect, and it will generally lead the children to study, with pleasure and wonder, the very creature of which they were positively afraid only an hour previous.

The Cuckoo Flower (Fig. 22)

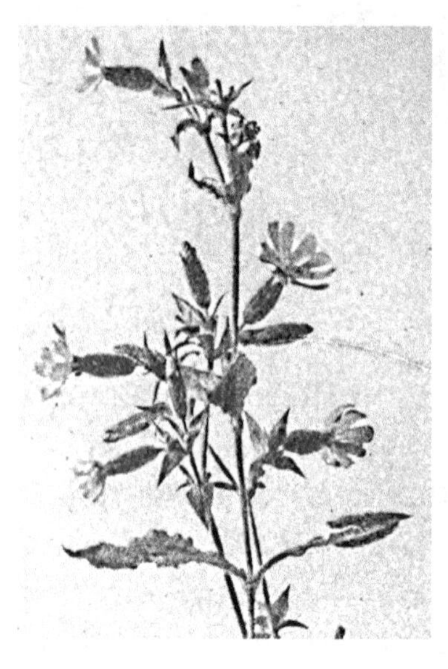

The Red Campion (Fig. 23)

The habits and life histories of many small animals can hardly be studied conveniently unless the animals are kept in captivity or semi-captivity for continuous observation; and we shall presently give examples of such, together with practical hints on the management of them. Some of the creatures to which we refer are of carnivorous, and even predatory habits, and would, therefore, appear to be quite unsuitable for the observation of children.

If predatory species are observed outdoors, we can know but little of them if we neglect to learn how they catch and devour their prey, for, in these instances, the most interesting points in their structure are those which fit the creatures for their predatory mode of life—the most interesting and instructive features of these (and of all other animals) are the wonderful adaptations of structure to habit.

Again, if we keep such animals in captivity, we must supply them with their natural food; and, in several cases, they will refuse everything but their living prey. The feeding of the creatures under observation, then, will be regarded as repulsive and cruel, and many teachers will unhesitatingly say that their habits are not such as should be witnessed by children, which means that children should be kept in profound ignorance of the true nature of the greater portion of the whole animal world—that they should never see a cat catch and eat a mouse or a sparrow, the spider catch a fly, or the lizard seize and devour a spider.

There is certainly a great deal of apparent cruelty in Nature, and we are not all agreed as to the advisability of doing our utmost to keep children in utter ignorance of the fact. Of course we must not attempt to make a display of the predatory habits of the animals referred to before children, especially young children; but, on the other hand, we must remember that, sooner or later, all children *will* learn that a very large proportion of the animal creation lives by preying on the more harmless species, and that by their own unaided observation.

The White Campion (Fig. 24)

It is for each teacher to decide whether he will allow his children to watch the habits of predatory creatures in the school aquarium and vivarium or not; but this much is certain—if the observations of animal life are to be confined exclusively to the harmless species, the children will get a very one-sided view of the works of Nature, and will lose the opportunity of studying some of the most marvelous examples of the adaptability of structures to the functions they have to perform.

The Herb Robert (Fig. 25)

Whatever be the animals studied, special attention should always be drawn to their movements, including the means of locomotion. In the case of the higher animals with well-developed limbs, the limbs may be compared with each other and with our limbs, both as regards their structure and their function. Those lower in the scale of life present a great variety in the form and movement of the appendages used for locomotion, the structure being always admirably adapted to the habits and habitats of the creatures; while in some instances, where special appendages for locomotion are either totally absent or are so minute that they are not easily observed, children will take considerable interest in trying to solve the problem of the creatures' movements from place to place.

The Sycamore in Flower (Fig. 26)

As previously hinted, the most interesting movements of animals are connected with the means of procuring and disposing of their food; and when children have an animal under observation they should be encouraged and guided by such questions as: What is its food? How does it find its food, or how does it catch it? Does the animal rely entirely on its strength and agility, or to a greater or less extent on stealthiness? Does it hunt for its prey, or simply lie in wait for it? Does it construct any kind of snare? How does it feed? Does it chew its food, swallow it whole, or does it feed only by suction? Does it store food not required for use at the time being? etc. Questions such as these, put to the children while observing or when about to observe, will help them to search more closely into things than they otherwise would.

Predatory animals are provided with the means of attacking and overpowering their prey, and the latter have, generally, some more or less effectual means of defense. Attention should be called to the various means of protection in those species that have enemies.

Some possess a kind of armor; others have weapons with which they can attack their foes. In the absence of both these, safety is often secured by burrowing, by the construction of some kind of home or shelter, by congregating together in vast numbers, or by a color-resemblance to environments.

Few things are more interesting in regard to the natural history of animals than this matter of resemblance to surroundings; and striking examples may be witnessed almost everywhere. Many town children will seldom, if ever, get the opportunity of noticing how various birds and mammals are assisted by such resemblances, and how, in some instances, the color even changes with the seasons in order to harmonize with the changed surroundings, but all may witness equally interesting cases among the lower animals.

The Comfrey (Fig. 27)

In town gardens and open spaces they may see the green plantbug, of the same tint as that of the leaf from which it is sucking the sap; the green caterpillar at rest on the projecting vein of the underside of a leaf, where it looks just like the vein itself; the twig-like caterpillar, with its straightened body standing out at an angle from the twig which it almost exactly resembles; and the moth, resting with depressed wings on a tree or fence on which it is scarcely discernible.

It should be noted, too, that resemblance to the environment is not confined to creatures which require protection from their enemies, for examples are common in which predatory animals are, by this means, enabled to lurk unseen in wait for their prey.

Furthermore, very interesting examples of mimicry are to be observed among animals. Some insects are protected from birds and other enemies by an obnoxious flavor, and

The Scarlet Pimpernel (Fig. 28)

others, though edible, by mimicking their color, are less liable to attack. A certain British fly that has no weapon by which to defend itself is seldom attacked because it so closely resembles a wasp; and if the fly is caught it will even imitate the movements made by a wasp when the latter tries to thrust its sting into its captor. Some children are shrewd enough to find out such facts for themselves; but, if not, the mention of a few instances by the teacher will make them keen in their search for similar examples.

The Oak in Flower (Fig. 29)

We have spoken briefly of the ways in which animals protect themselves from their enemies, but equally instructive are the methods by which some of them secure shelter from unfavorable conditions of the weather, and the preparations made by others previous to the assumption of a dormant or passive state. Examples of such will be mentioned when dealing with the animals concerned.

Children should also be encouraged to observe the various ways in which animals provide for their offspring: whether they lay eggs and, if so, when and where; whether any kind of provision is made for the safety of the eggs and of the young that are hatched from them. If any kind of nest is made, they will study the nature of the nest, and, if possible, watch the progress of the building. If no nest, they will note whether the eggs are placed in such a position that they are protected from extremes of temperature and from other dangers.

Some animals leave their eggs entirely as soon as they have been deposited, but it will be noted in such instances that the eggs are generally left in some safe situation, and that they are laid either on or near the very food which the young ones require.

If the eggs are cared for by the parent, then the following matters should be investigated by the young observers: How are the eggs protected and cared for? Are the young ones fed; and, if so, how are they fed, and with what food?

Furthermore, where the same animal has been kept under observation for some time, the children may have the opportunity of ascertaining the length of its natural life.

II. Some lower forms of animal life

Almost all kinds of animals may be studied during the spring as soon as the sun is sufficiently warm to call the hibernators from their winter quarters, but some are especially suited for observation during this period.

Snails and slugs are now out of their hiding places, and during damp weather, especially in the evening after the sun has gone down, they may be found feeding on the early vegetation.

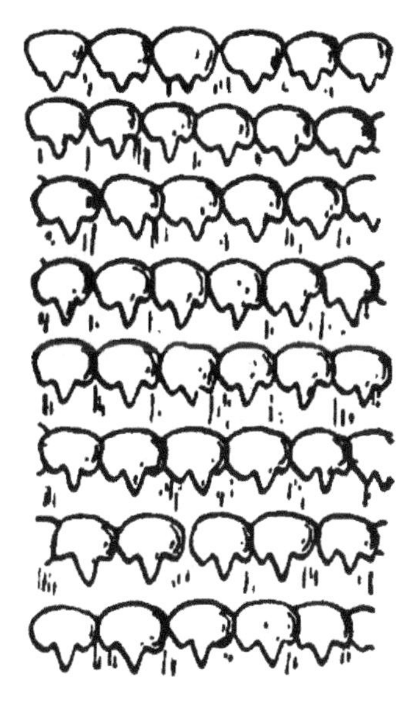

The teeth of a snail (Fig. 30)

A few of these may be kept for a time in a box with a glass front, covered with a lid of perforated zinc or other suitable material. They will eat almost any kind of vegetation that we ourselves find edible, and the manner of feeding may be observed with the aid of a lens.

If a hungry snail be placed in a rather deep but small jar or box, and provided with a piece of cabbage leaf, the rasping action of its minute teeth may be distinctly heard as it feeds, by placing the ear close to the mouth of the box. In order to see its wave-like muscular contraction of the lower surface of its body as it creeps, we place the snail on a piece of glass, and look at it through the glass.

Among the other numerous interesting features of the snail we may briefly note the retractile, sensitive 'horns', which are pulled outside-in by means of small muscles when touched; the eyes at the tips of the longer 'horns'; the breathing hole under the rim of the shell on the right side of the snail's body;

and the out-turned rim of the shell of the adult snail, which shows that the growth is now completed.

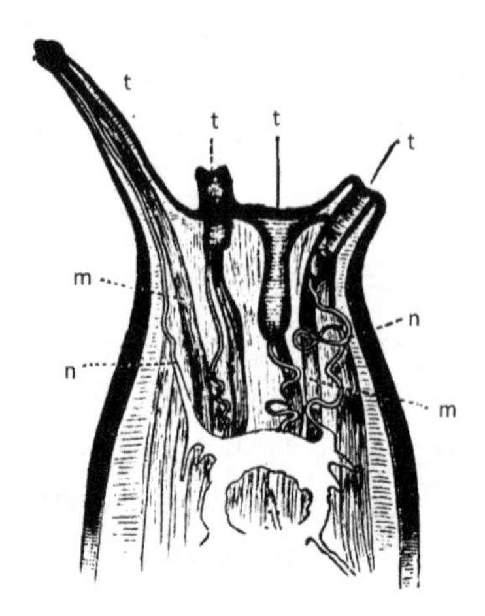

The head of a snail, showing horns (t), the muscles in them (m), and the nerves (n) (Fig. 31)

Of course the children will note many other interesting features concerning the structure and movements of the snail, and also of the slug, which is really only a species of the same group with no shell or only a rudimentary one. We merely call attention to a few points that may possibly be overlooked in the absence of simple hints and directions.

Several very interesting species of water snails are to be dredged from weedy ponds, during the spring, with the aid of a muslin or gauze net; and if some of these are transferred to a glass aquarium, together with some pondweed, their movements may be observed with ease.

If this is done, it will not be long before we notice cylindrical masses of jelly-like substance, containing minute specks, attached to the pondweed or to the glass. These are the eggs of the water snails; and if we examine them from time to time, we can observe the gradual development of the young snails. The jelly-like covering of the eggs is quite transparent, and those masses of eggs that are attached to the glass sides of the aquarium may be closely examined through a magnifying lens.

An important difference between land snails and water snails must be noted. The former take their supply of oxygen direct from the atmosphere, through the breathing aperture previously mentioned, the organ of respiration being a lung with a function similar to that of our own lungs; the latter, on the other hand, breathe by means of gills which absorb dissolved oxygen from the water in which they live.

Earthworms are also very interesting animals for study. Encourage the children to observe them in their natural haunts. They are of nocturnal habits, coming to the surface after darkness has set in, especially when the soil is wet, but easily obtained in the daytime by digging. They swallow the soil as they burrow, digesting from it the decomposing vegetable matter that forms their food, and this soil is ejected at the surface in the form of little wormlike mounds known as worm-castings. Thus they perform a very useful work, for their burrows assist in the aeration of the soil; and in continually bringing the deeper soil to the surface they are doing a work similar to that of the plough.

The attention of the children should be drawn to the general cylindrical form of the body, the flattened hindmost portion, the large number of ring-like segments, the absence of limbs, and, of course, the movements.

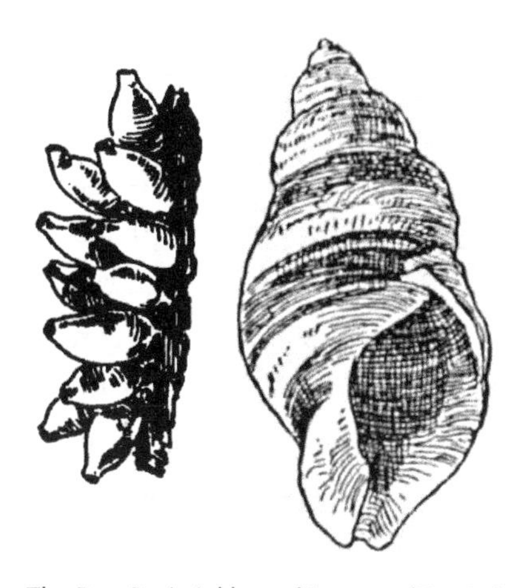

The Dog Periwinkle and its eggs (Fig. 32)

The Dog Whelk (Fig. 33)

By stroking the body of the worm the wrong way we can feel the little bristles, of which there are eight on most of the segments, all directed backward. These are used for locomotion; the worm extends the fore portion of its body, and then, applying its bristles against the ground, is enabled to pull forward the hind portion by muscular contraction.

If it is desired to keep a few earthworms in the schoolroom for closer observation, put them in a wooden or other suitable box half filled with a soil that is rich in leaf-mold. Keep this soil wet and cover the box with a glass plate. Another and better method of keeping worms for observation will be found on page 213.

In the case of schools situated within easy distance of the sea the children may be stimulated to observe the forms and habits of the marine snails commonly known as whelks and periwinkles, and the marine worms.

The former include the common whelk, dog whelk, common periwinkle, dog periwinkle, sting winkle, and the limpet—all of which are abundant on every part of our coast. The first of these really lives in deeper water some distance out, but may always be obtained alive, when required, from fishermen; but the others are always to be found on rocks and stones at low tide. With the exception of the limpet, which has a conical shell, all have spiral shells and, of course, spiral bodies; and all of them breathe by means of gills like the fresh-water snails.

During the spring months the pretty stalked eggs of the dog periwinkle may be found attached to the rocks, usually in crevices and under the shelter of projecting ledges; and the eggs of the large common whelk are often washed on the beach in large numbers by the waves, sometimes still unhatched, but more frequently only the empty cases.

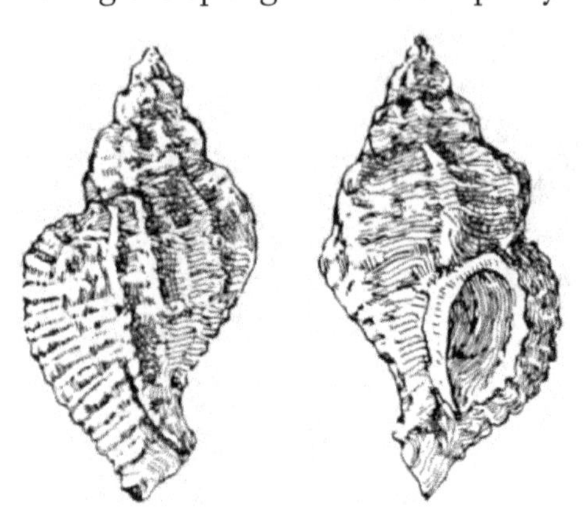

The Sting Winkle (Fig. 34)

The movements of the animals named may be conveniently observed in a glass aquarium supplied with fresh sea-water, and the development of the young from the eggs may also be watched.

Eggs of the Whelk (Fig. 35)

Marine worms are even more interesting than the earthworm, and some of them are really beautiful objects. Several species burrow into sand or mud at the seaside, and these may be obtained by digging, and observed in a salt-water aquarium. Some construct tubes of sand, the particles of which are bound together by a secretion from the body, while others live in hard tubes of limy material that are usually attached to weeds, shells and stones. The burrowing species often leave their burrows and swim freely, but the tubebuilders lead a sedentary life.

The beauty and the interesting movements of these marine worms are best observed in the aquarium, for they seldom expose themselves when out of the water. Unlike the earthworm, they have very pretty gills and other appendages, but they are all similar in having long, segmented bodies.

Seaside observers should also study the beautiful anemones—variously-colored jellyfishes, with cylindrical bodies and numerous feelers, that live attached to rocks, stones, shells, etc. These should also be observed in the aquarium, or in the rock-pool, for it is only in water that they expand their beautiful bodies and feelers. They may be kept alive a long time in captivity if supplied with fresh sea-water at intervals, and fed with little pieces of fish.

A common sea anemone (Fig. 37)

It should be noted that the sea-anemones are much lower in the scale of life than worms and snails. The latter are very complicated in structure internally, having well-formed digestive tubes, digestive glands, blood system, excretory organs, nervous system, etc.; but the anemones are very simple animals, with no internal organs in the ordinary meaning of the term, and the only digestive apparatus is a portion of the general body-cavity, partly shut off from the remainder, that serves the purpose of a stomach.

Finally, it may be mentioned here that both fresh-water and salt-water pools are especially interesting in the spring season, not only on account of the objects already named as inhabiting them, but because they now contain a variety of animals that have peopled them afresh after their winter hibernations, together with many others that are now in their early, immature stages. Some of these belong to higher forms of life than those we have been considering, and will be mentioned in their proper places.

Marine tube-building worms (Fig. 36)

Nevertheless, it will be useful, at this time, to search the ponds and pools with a collecting net, and thus procure a stock of interesting material for the school aquaria, the management of which is dealt with in Chapter 13.

III. Spiders and centipedes

Passing now to animals that are more highly organized than worms and snails, we shall first make a few observations on spiders and centipedes. These, together with insects and crustaceans, form the large and important group called the arthropods, the chief characteristic of which is a segmented body bearing jointed limbs and other appendages.

Both spiders and centipedes are, perhaps, best studied in their natural haunts outdoors; but they may, of course, be kept for a time in captivity for closer and more continuous observation.

There are a large number of British spiders, varying much in size, general appearance and habits, but we specially recommend the very common garden spider, sometimes called the cross spider on account of the light, crosslike mark on the top of its body, as a very interesting one for study, and very easily found.

It is common among children, and even among adults, to look upon all small creeping things as insects. But spiders belong to a group of the arthropods quite distinct from insects; and when the former are being examined, particular attention should be called to those features in which they differ from the latter. Such features will be noted as we proceed.

For the guidance of the teacher we give a brief summary of the main characteristics of the garden spider. First, we observe that its body is made up of two distinct parts (not three, as in insects), the foremost being a combined head and thorax, and the hinder part the abdomen. To the front portion are attached four pairs (insects have only three pairs) of jointed legs, covered with stiff hairs, and terminating in several claws; and the large, rounded abdomen has no appendages, and is not segmented as is the case with insects.

Attached to the head is a pair of long, jointed feelers; a pair of forceps (*falces*) above the mouth, by means of which the creature captures its prey and injects poison into its body; a pair of jaws, one on each side of the mouth; and eight minute eyes.

Under the abdomen, towards the front, are two slits through which the spider breathes; and, behind, the little silk-spinning organs, two of which are longer than the others and project.

Of course, some of the above organs are not to be seen except on a very close examination, but all are revealed if we employ a good magnifying lens to study a spider that is temporarily imprisoned in a small glass-topped box.

As regards the habits of the spider, the most interesting are undoubtedly the spinning of the beautiful web, and the capture and disposal of the prey. The construction of the snare may be watched in the open, but the spider will also perform the task indoors, if preferred. Should the latter be desired, take a garden spider from its web, put it in a large glass case containing a few twigs to form a support for the new snare, and place the whole in a well-lighted spot.

The method of procedure is as follows: First the creature lays a horizontal thread from one point to another, often making use of the wind as a means of carrying the thread across, but sometimes climbing across and putting the thread in the desired place. The newly-spun thread is still somewhat sticky, and adheres firmly wherever it is applied. Several other threads are now formed, one by one, each fastened to the first, and then carried in various directions to suitable points of support, thus completing the framework of the future structure.

The spider next comes to the center of the horizontal thread, and, making many somewhat similar journeys, spins a number of threads that radiate in all directions from the middle to points on the outer framework. So far, all the threads spun by the liquid silk dry rather rapidly, losing their stickiness, and, therefore, do not form part of the actual snare. This latter consists of a continuous thread, passing from radius to radius, forming a fairly regular spiral. In constructing it the spider starts at the center, and gradually widens the circumference of its circular journeys as it proceeds; and this thread is of a different nature from the rest of the web, since it supports numerous minute globules of a liquid silk that does not harden on exposure to air, and therefore forms the snare for the capture of insects.

Although so very fine, each thread of the web is really compound, being formed by the combination of all the liquid threads that issue from the numerous holes of the spinnerets.

It will be observed that the garden spider is very variable in color, but it may always be recognized by the more or less distinct cross on the top of the abdomen. The male is not nearly so large as the female, nor is he so fierce and voracious as she.

The female lays her eggs in the autumn, usually in October, and covers them with a roundish, yellow, silken cocoon, about two-thirds of an inch in diameter. These cocoons may be found on garden fences and the trunks of trees all through the winter. The young ones, several hundreds in number, hatch out in the following spring; and, immediately on leaving the cocoon, they spin a common web, inside which they live for a short time, all clustered together into a little ball about a third of an inch in diameter. In this condition they may be found suspended from some object near their winter nest. Touch the ball of young spiders, and immediately they dart asunder on their protecting web; but as soon as the alarm is over they resume their former crowded condition. It is not long before the young spiders separate from one another, and then each one constructs its first tiny snare for the capture of the small insects that now form its food.

Finally, it must be noted that the newly hatched spiders are of the same form as the parent. They do not undergo metamorphosis, in which respect they differ from the majority of insects, as will be seen presently.

We need hardly remind the teacher that he should not find it necessary to impart much of the above information to his class, but that the children should, as far as possible, make the whole a matter of original research. The information is given to enable the teacher to augment or correct his own observations, if necessary, before putting the children on the same work; and the same remark applies equally well to all other information given in this guide.

It will be seen, too, from our remarks on the spider, that the cocoons should be searched for in winter or early spring, and the young spiders later in the spring, while the mature creatures will provide material for study during the summer months.

Again, these remarks apply particularly to the common garden spider, but it must be remembered that there is a large number of British species, all differing from the others in some special features; and thus, if observations are made of other species than the one we have referred to, differences must be expected. Some species, for instance, have bodies of quite a different shape;

some have only six eyes; some females carry their cocoons attached to their bodies; others construct no snares, etc.

We can hardly close our account of the spiders without a brief reference to the very interesting water spider. This creature is common in ponds and slow streams, and may be readily obtained with the aid of a muslin or gauze net from April onwards. Its wonderful habits may be observed if it is placed in an aquarium containing a little pondweed of any kind. It lives under water, but derives the air necessary for respiration direct from the atmosphere. Its abdomen is covered with short hair, between which is always entangled a supply of air that gives the creature somewhat the appearance of a moving ball of silver. When the oxygen requires replenishing the spider comes to the top, and pushing its abdomen above the surface of the water, disengages the vitiated air, replacing it by a fresh supply.

Water spiders and their cells (Fig. 38)

The water spider also constructs a silken home, in the form of an inverted, oval cell, fastening it by means of silk threads to the water-weeds; and this cell is filled with air in the following manner. The spider comes to the surface, pushes the end of its body into the air, and then, by a rapid movement of the hindmost logs, captures a bubble of air between the legs and the spinnerets, carries it down and releases it in the cell. This process is repeated until the cell is nearly full of air, when the spider will get inside, take up a position with its head downwards, and watch for its prey.

Centipedes are not so interesting as spiders, but they are worthy of a little study if only because they represent a distinct division of the jointed animals. Their bodies consist of a chain of similar segments, each bearing a pair of short, jointed limbs. They are nocturnal creatures, hiding under stones, in the soil, in rotten wood, or in some other situation where they can avoid the light during the day, and wandering in search of their prey after dark. They are very voracious, are armed with a pair of powerful jaws, and often attack earthworms and grubs much larger than themselves.

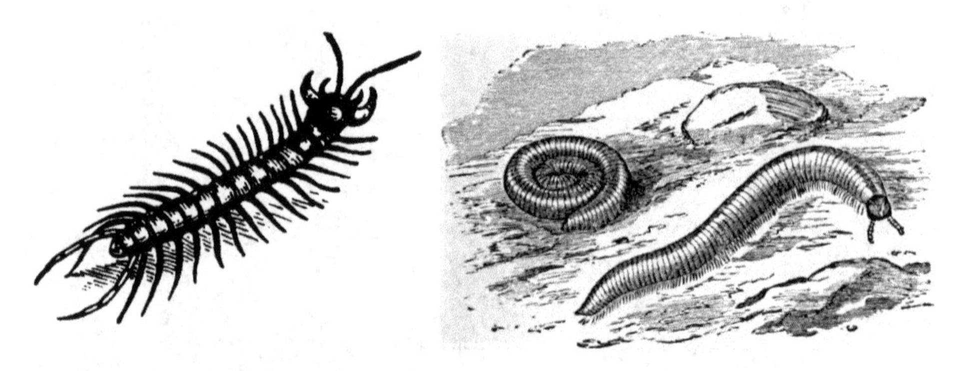

A centipede (Fig. 39) *A millipede (Fig. 40)*

Millipedes are somewhat similar in general structure and habits, but their bodies are made up of a much larger number of segments, with a corresponding number of pairs of legs. They are not all carnivorous (for there are several species), some feeding on decaying organic matter, while a few are said to attack living plants. They roll up into a spiral when disturbed.

IV. Insects

During winter and early spring hardly a sign of insect life is to be seen except by those who know where to search out the hibernating species; but as the latter season advances insects appear on the wing in gradually increasing numbers, affording abundant opportunities for the most interesting observations.

Most insects are very unlike their parents in their early stages, and pass through marvelous changes before their mature condition is attained. In the higher forms there are generally three distinct stages—the larva or grub, with or without legs; the pupa or nymph, sometimes quite quiescent or passive; and the imago or perfect insect, usually possessing two or four wings and endowed with enormous power of flight. The eggs are laid only during the last or perfect stage, and from these are hatched the larvae or grubs of the future generation.

All insects must, of course, pass through the winter in one of the above stages. In some species the eggs are laid in the autumn and remain unhatched until the following spring. Others hibernate in the larval condition, seeking shelter from the winter weather under the soil or under some other cover. Others, again, spend the whole of the winter and early spring in a quiescent pupal state; while a smaller proportion hibernate in the perfect or winged condition.

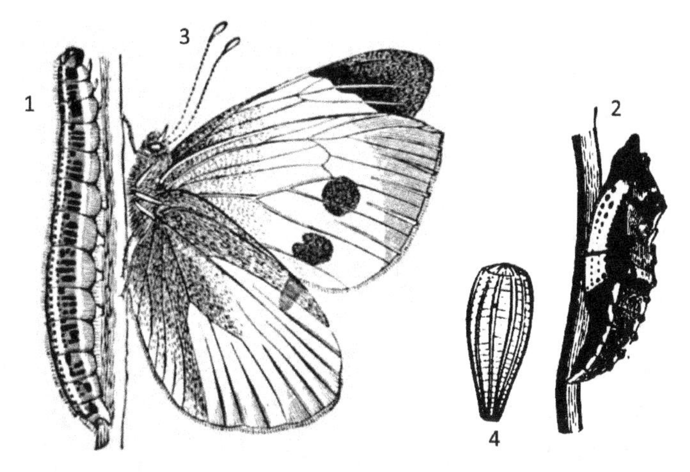

1. Larva or Caterpillar, 2. Pupa or Chrysalis, 3. Perfect insect, 4. Egg (Fig. 41)

The study of insect life should not be totally neglected during the cold months of the year, for it is interesting to learn the various ways in which insects protect themselves from harm during this period.

When preparing the soil for the spring flowers and crops, we are sure to meet with many of them—brown or black pupae, covered with a hardened, protecting skin, snugly housed in a cozy little oval cell scooped out of the soil, with walls often made of particles of earth bound together by threads of silk; pupae or chrysalides similarly housed and protected, frequently in a neat silken cocoon, or in a hard shell formed by gluing together earth or other material with a silky secretion, often with the additional shelter afforded by the angles of the roots of the trees or shrubs on which the insects fed; and larvae and pupae lying on the surface of the soil, under the cover of fallen leaves that are sometimes bound together to form a suitable winter home.

Hibernating insects, in various stages, may also be found on fences, in the crevices of the bark of trees, and in holes and under protecting ledges almost everywhere.

The eggs and larvae found during this season may be collected and reared with the object of studying the future life histories, providing the species and the food required are known; but in all cases the pupae may be preserved in order that the perfect insects may be seen and, if possible, the actual emergence of the same be observed.

Whenever hibernating larvae or pupae are collected for future observation, they should be kept as nearly as possible under their natural conditions until the period of hibernation is over. Those found in the soil must be kept in soil in which a certain amount of dampness is preserved, while those obtained from dry situations simply require housing in a suitable box placed in a cool, airy situation.

Butterflies that hibernate (Fig. 42)

1. Brimstone, 2. Peacock, 3. Small Tortoiseshell
4. Comma, 5. Red Admiral, 6. Painted Lady
7. Large Tortoiseshell

From October to about the end of March the angular chrysalides of common white butterflies—the large white and the small white butterflies—may be found under the projecting portions of garden walls and fences, both in town and in country. These are generally placed horizontally, and secured by a silken tuft at the tapering, hind extremity as well as by a fine but strong silken cord around the middle. Some of these should be secured in order that the school children may have the opportunity of observing the emergence of the butterflies, followed by the rapid expansion and subsequent stiffening of the wings, the two species named being admirably adapted to this purpose.

To remove the chrysalis without injury, first cut the silken thread that supports the body at about the middle with a pair of pointed scissors, so that the chrysalis now hangs vertically by the 'tail'; and then carefully scrape away the silky tuft at the tail from the surface that supports it.

April Butterflies (Fig. 43)

1. Green-veined White, 2. Small White

3. Large White, 4. Orange Tip, 5. Do (underside)

6. Wood Argus, 7. Small Copper

To increase the chances of witnessing the interesting event referred to above, several of the chrysalides should be procured, and preserved in a cool place till about the end of March. About this time, but varying more or less according to the temperature of the season, a gradual change may be observed in the color of the chrysalides—the white color of the developing butterfly, and especially of the wings, becoming more and more distinctly visible through the semi-transparent case.

On a warm, sunny day in spring, when several of the chrysalides show, by the appearance just referred to, that they are ready for the final change, place them in direct sunshine, in the center of a group of children, and the chances are very great that the children will witness one of the most marvelous episodes of insect life.

On mild, sunny days in early spring, and on similar days even in the midst of winter, certain beautiful butterflies may be seen on the wing, searching out the few early flowers that have been encouraged to expand their petals by the warm rays of the sun. These are not early butterflies of the present new year, but perfect insects that have hibernated in sheltered places since the previous autumn. Later in the spring, when the weather is warmer, they may be seen on every sunny day, and their life ends soon after the deposition of the eggs that produce the caterpillars of the future generation. The commonest of these hibernating butterflies are the brimstone, tortoiseshell, peacock, red admiral, and painted lady butterflies.

Several species of moths also spend the whole of the winter in the perfect state; but since these fly only at dusk or after dark they are not so commonly seen. They may, however, often be observed during the day, at rest on fences, the bark of trees, etc.

Other butterflies and moths which have spent a portion or the whole of the winter in the chrysalis state soon make their appearance. The earliest of the moths may be seen in flight even during January and February, but the first of the butterflies—the large white, small white, green-veined white, orange-tip, wood argus, and the small copper butterflies—do not usually make their appearance till the middle or the end of April.

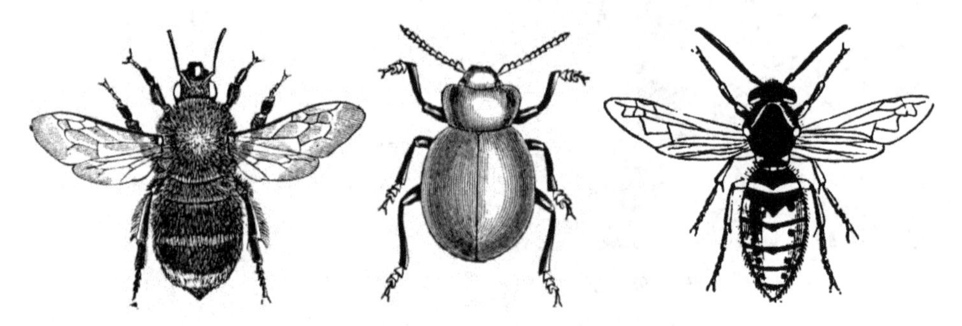

The Queen Humble Bee (Fig. 45) *The Bloody-nosed Beetle (Fig. 46)* *The Queen Wasp (Fig. 44)*

As the spring advances the number and variety of insects are rapidly on the increase. Queen bees and queen wasps, the only survivors of the broods of the previous year, may be observed searching out a suitable place for the construction of a new nest and the establishment of a new brood or colony beetle, commonly known as the bloody-nosed beetle on account of the red fluid it ejects from its mouth when disturbed, is now abundant on hedgerows and banks. Ants become very busy in the construction of new nests and in the enlargement of old ones to accommodate the rapidly increasing colonies.

Various two-winged flies are feeding on the nectar of flowers and, in the case of some species, on decomposing organic matter, by the odor of which they have been attracted. Among the nectar-feeders perhaps none are more interesting than the humble-bee fly, which closely resembles a bee, and which hovers over the early flowers, and sips the nectar through its long proboscis while poised on the wing.

At the same time the ponds and streams are being rapidly repeopled with insect life. Quite a large number of grubs—the larvae of gnats, dragon flies, caddis flies, may flies, water beetles, and of various other aquatic insects may be obtained in numbers by means of a muslin net. These, if placed in the school aquarium, and treated according to the advice given on page 205, will give splendid opportunities of watching numerous examples of insect metamorphosis. The gnat larvae, which may be found in almost any still pool or open water-butt, are very suitable for such observations, for they pass through their stages so rapidly that, with a large number in stock, the changes may be witnessed at almost any time.

V. Fishes

Fishes may be studied almost equally well at all times of the year, but we think it advisable to give some general hints concerning them now, since they are most conveniently observed in the aquarium and since the spring season is the best in which to stock aquaria, with material for the summer's study. The method of starting and maintaining aquaria will be found in Chapter 13, and we shall deal here only with the principal features that should be noted when the fishes are under observation.

Hitherto we have spoken only of invertebrates, but in the fishes we become acquainted with the lowest of the vertebrate or back-boned animals. They have, at least generally, a bony skeleton, including a back-bone, skull, ribs, and imperfectly-developed limbs.

The fish most generally selected for school study is the goldfish, chosen, probably, partly on account of its pretty coloring, and partly because it may usually be obtained from a dealer without the slightest trouble. We doubt, however, whether it is the best for our purpose. Its movements are very graceful as well as instructive, but they soon become somewhat monotonous.

This is not the case with the interesting little stickleback, which is to be seen in nearly every pond and stream. This fish is easily procured by means of a muslin net on the end of a stick. It is so hardy that it is easily kept alive and healthy in an aquarium, providing it has plenty of room and the proper food; and its habits and movements are so varied and interesting that for a considerable time we can be learning something new concerning it.

During the springtime we meet with what are apparently two kinds of sticklebacks in fresh-water ponds, but these are really the two sexes of the same species. At this season, and at this season only, the male stickleback is very brightly colored, the chief distinguishing mark being the large patch of bright red that has earned for him the name of 'red-throat.'

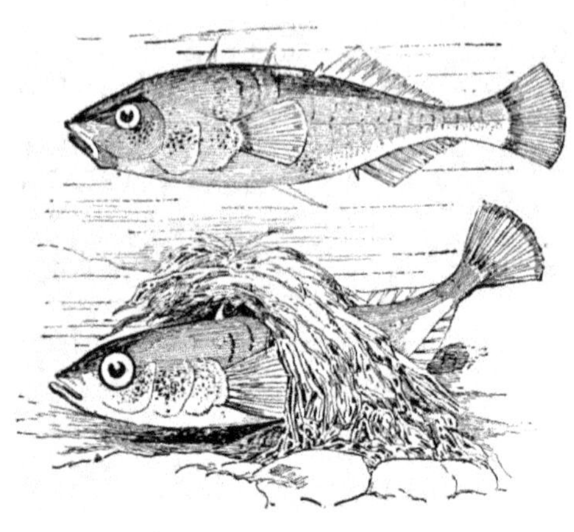

The Stickleback and its nest (Fig. 47)

This season, too, is that during which this interesting little fish is concerned in the matters of home and family. The male builds a nest by weaving together the stems and leaves of pondweeds—a kind of tubular nest, very loosely put together by the aid of the creature's mouth. The female deposits her eggs within this nest, and during the whole time that the nest is required—that is, until the young ones appear—the male guards it most carefully against enemies of all kinds, including those of his own species.

During all this period the male stickleback is most vicious in his treatment of intruders that approach too near his domain. He will furiously repel all intruders, regardless of their size; and if threatened with a walking-stick, he will charge the stick with such fury that, although he is so small, the blow will be felt at the other end.

The stickleback is at all times voracious in his appetite for worms and grubs, generally pugnacious and quarrelsome in his habits, and always full of vigor. With such qualities in addition to its nest-building and house-guarding habits and all the other features common to fishes generally, it will be seen that the stickleback must be a most interesting fish for school observation.

As regards the general structure and habits of fishes, children should be encouraged to note that the shape of the body is admirably adapted for easy progression through water; that a fish is, with the exception of the heavier bottom fishes, of about the same density as the water in which it swims; that its body is covered with smooth, overlapping scales with their free edges turned backwards; and that it is apparently always drinking, but in reality is only taking water to bathe the gills for respiratory purposes, the water passing out under the gill-covers after it has been deprived of some of its dissolved oxygen.

It should also be noted that there are (in most fishes) two pairs of fins, corresponding with our own limbs, in addition to other fins in the medial plane of the body—on the back, on the ventral surface, and at the tip of the tail. Concerning the uses of these fins for progression, balancing and steering, the children must get their own ideas of them by carefully watching the movements of the body of the fish resulting from the different motions of the fins. The children may be wrong sometimes in their conclusions, but better this, after some real attempt, than to be told by the teacher and thus deprived of all stimulus to thought.

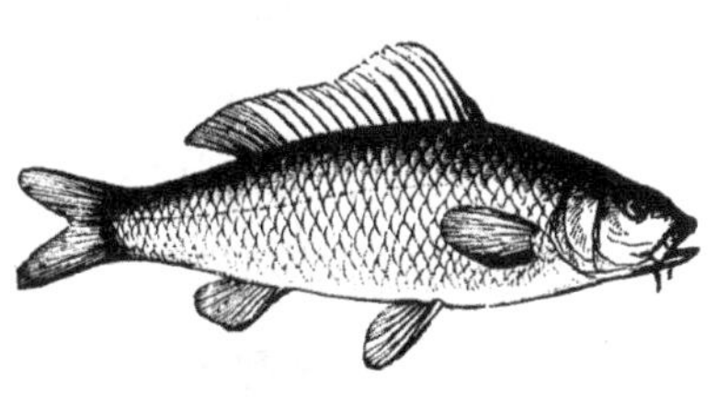

The Carp (Fig. 48)

Perhaps the commonest of all mistakes made as regards the function of the fins of fishes is to conclude that the tail, because it is situated in a position which corresponds with the rudder of a boat, is used solely for steering purposes. A little careful observation, however, will show that the tail fin is really the principal propelling organ of the fish.

Of course there are several other common fishes in pools and rivers that are suitable for school study, such as the carp, roach, minnow, gudgeon, loach, etc. The last of these is one to be recommended, especially because its habits and movements, as a bottom fish, differ from those of most familiar fresh-water species.

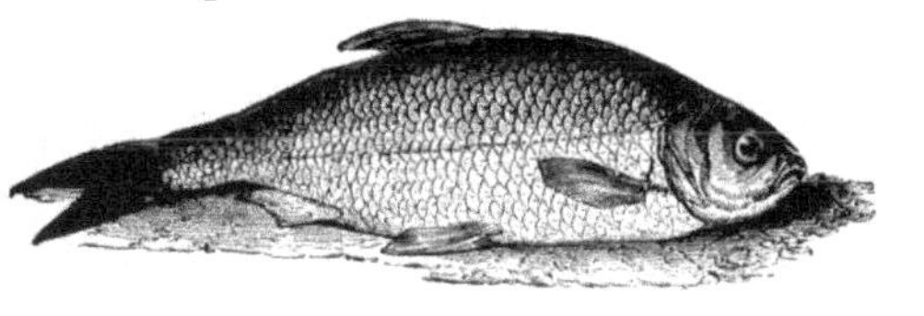

The Roach (Fig. 49)

The Loach (Fig. 50)

The body of the loach is heavy, as with bottom fishes generally, and the fish spends much of its time at rest on the bed of the pond in which it lives.

It is provided with feelers (barbules) by means of which it can feel its way about on the ground, and is naturally of a less active nature than the free-swimming species. Placed in an aquarium with water lilies or other plants that have floating leaves, it will often come to the surface, rest on a leaf, and even bask in the sunshine with its body half out of the water; and in this position, it will, after only a short period of captivity, take its food (small worms, etc.) from the hand.

The Minnow (Fig. 51)

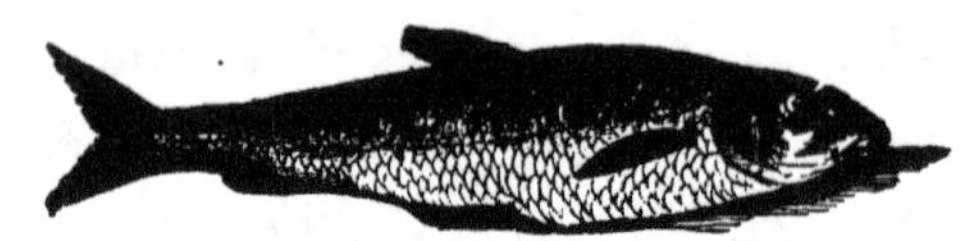

The Gudgeon (Fig. 52)

Although we have referred to two fishes in particular as being suitable and interesting for school study, we do not wish it to be understood that these are all-sufficient. Fishes, like other divisions of animal life, include a very varied collection of creatures, and it will be well to make ourselves acquainted with the greatest possible variety of structure and habit.

The Butter-fish (Fig. 53)

The Smooth Blenny (Fig. 54)

The children of schools in seaside towns and villages would naturally be led to study some of the common fishes of the sea; and while a certain amount of time may be devoted to the observation of the leading characteristics of form and color by which we distinguish between the various food fishes brought in by the fishermen, a much more valuable work, from an educational point of view, may be accomplished by the study of the habits and adaptations of the little fishes that are to be seen alive under stones and in rook-pools left by the receding tide.

Let the children ramble on the shore when the tide is low, and turn over stones near the water's edge. Here they will find the slippery little butter-fish, the bony and sluggish pipe-fish, the pugnacious smooth blenny, the equally pugnacious marine bullhead, and other small species with most interesting habits and remarkable adaptations of structure to habit.

The Marine Bullhead (Fig. 55)

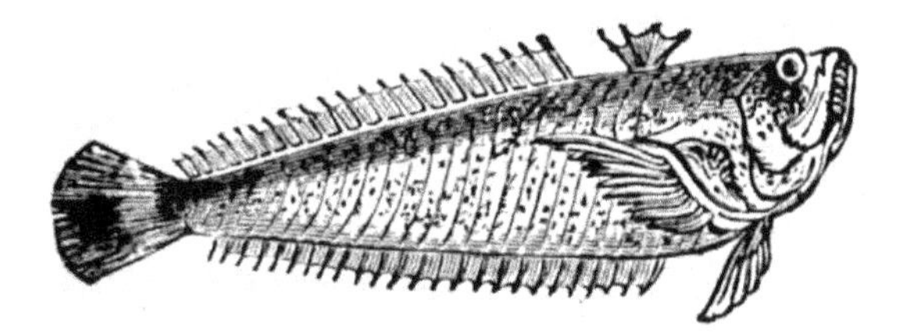

The Little Weaver Fish (Fig. 56)

These little fishes will live many hours, and some of them even days, out of water, providing they are kept wet with sea-water. Their movements may be closely observed by transferring them to a little rock-pool, or they may be taken home in wet seaweed, and then placed in a salt-water aquarium furnished with some weeds and stones to provide them with suitable cover.

Then, in the rock-pools themselves, we have beautiful little natural aquaria, ornamented with weeds of various colors, and harboring little fishes of various species, including those we have previously named. Here the fishes are to be studied under perfectly natural conditions, and with as much convenience as when they are housed in a glass aquarium.

Even on sandy shores, where stones and rock-pools are absent, there are favorable opportunities of observing certain species of fishes that come close up to the water's edge. On a calm day, when the water is undisturbed except by the passage of little ripples, young flat-fish, with their upper sides mottled in such a manner as to be scarcely distinguishable from the sand itself, may be seen swimming from place to place in the shallow water; and, after each short journey, settling on the sand, and suddenly hiding themselves by throwing up sand with the aid of their fins to cover their bodies.

Skate eggs (Fig. 57)

The troublesome little weaver fish, too, will bury itself in sand, leaving nothing exposed except its eyes and the spiny weapon on its back, the latter often causing considerable pain by piercing the feet of unwary little paddlers.

In addition to these and other fishes that may be searched out on the sea shore, there are also interesting objects connected with fish life to be seen among the debris washed up on the beach. Thus, clusters of the eggs of fishes are frequently to be seen cast up on the sand, or attached to seaweeds or to the rocks. If these are placed in a salt-water aquarium they will hatch out, giving an opportunity of learning that the young fishes come into the world in a very imperfectly developed condition—that they have to pass through a larval condition, like frogs, toads and newts, before they finally assume the form of the parent.

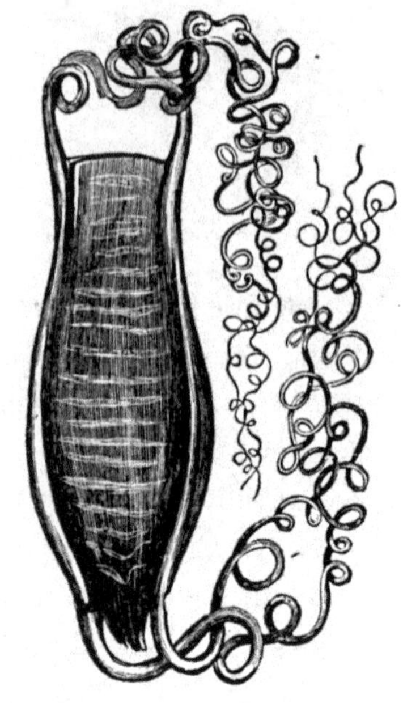

The egg of the Dog-fish (Fig. 58)

The large, horny eggs of the skate and the dog-fish are also to be found among the thrown-up debris at high-water mark. More frequently the empty egg-cases of these fishes are cast up after the young have hatched out, and a slit at one end then shows where the little fish made its escape; but occasionally the unhatched egg is washed up intact, in which case the young one may be seen snugly coiled up within its cell when viewed by looking through the latter in a good light, and the hatching may then be as before described.

VI. Amphibians

We now come to the amphibians—those animals which spend the early part of their lives as fishlike larvae or tadpoles, breathing by means of gills, and then develop into four-legged creatures that breathe by lungs and are more or less terrestrial in their habits.

This division of animals includes the frog, two species of toads (the common toad and the natterjack toad), and three species of newts (the common newt, the palmate newt, and the great warty newt).

The study of these animals is full of interest, not only on account of the structure and habits of the adult creatures, but more particularly because of the remarkable changes through which the young have to pass before they assume the perfect form. In the latter state they may be observed at any time except during the winter months, when they are in a dormant condition; but the study of their metamorphosis is essentially an occupation for the spring.

The Frog (Fig. 59)

Frog eggs are to be found in almost every pond during the latter portion of March or in early April, the exact period varying according to the temperature of the season. They should be procured as soon as possible after they are laid, and placed in an aquarium with a liberal supply of any common pondweed. For some weeks, while the earlier stages of development are in progress, they will require no particular treatment, the pondweed supplying the young tadpoles with all they need in the way of food, as well as with the oxygen necessary for respiration; but after a time they will require a change of food and changed conditions, particulars of which will be found in our chapter on the management of school aquaria.

The study of the life history of the frog should consist of a series of observations carried out at fairly regular intervals for a period of a few months. While these observations are in progress the teacher should never attempt to assist the children by giving information as to future events and changes, but require the whole study to be one of original research on the part of the class, merely offering explanations in the case of problems which the children cannot work out for themselves.

Even young children should be encouraged to keep a complete record of the observations, with dates, illustrated by means of sketches drawn as near as possible to the natural size. The aquarium should be viewed about three or four times each week, at least during the early stages, but each observation need not occupy more than a few minutes.

The Common Toad (Fig. 60)

The development of the toad may be observed in exactly the same way, and the results compared with the corresponding stages of the frog. The eggs of the latter are always in large rounded masses of some hundreds, while those of the toad are embedded in long strings of jelly, and are usually laid a little later than the eggs of the frog.

The egg-chain of a toad (Fig. 61)

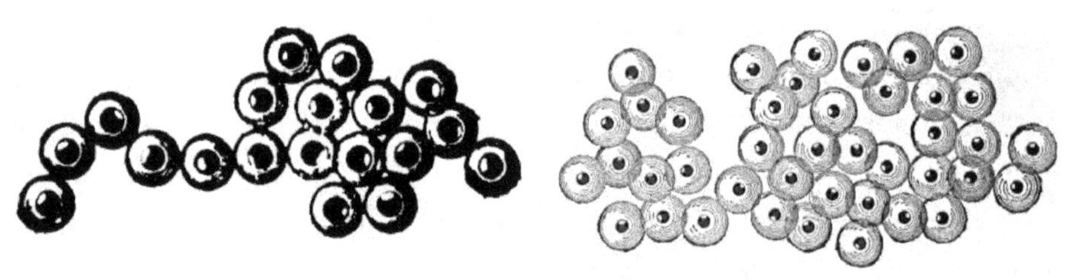

Frog eggs when they are first laid, and how they appear a little later (Fig. 62 & 63)

As a guide to the teacher we give a brief summary of the changes to be observed in the case of both the frog and the toad, and in the proper order:

1. **The Eggs**
    a) When first laid, the eggs consist of a black embryo surrounded by a rather thin layer of gelatinous substance.

b) The gelatinous covering soon absorbs much water and swells enormously, so that the mass of eggs laid by a single frog becomes very large.
c) In a few days, the tune varying according to the temperature, the embryo becomes elongated.
d) A few days later a little fish-like body may be seen moving within each gelatinous sphere.
e) The little fishlike tadpoles escape from the eggs, and cluster around the outside of the gelatinous mass.

2. The Tadpoles

a) At first these have no mouths, but are still supported by the remainder of the 'yolk-food' within their bodies; and they cling to external objects by means of a pair of suckers on the undersurface of the head. They breathe by branching, tufted gills at the sides of the neck.
b) A few days later the mouth is formed, and the tadpole commences to feed on vegetable matter, secured by the aid of its horny jaws.

Tadpole development stages (Fig. 64 & 65)

c) Four slits are formed on each side of the neck. The margins of these slits become folded, and form internal gills, while, at the same time, the external gills gradually decrease and finally disappear.
d) A fold of the skin covers the gill-clefts on each side, but leaving an opening for the escape of water on the left side only. Water is now taken into the mouth for respiratory purposes, and this, after bathing the gills on both sides, and giving up dissolved oxygen to the blood, passes out through the opening on the left side.

e) The tadpole is now much larger, and the hind legs begin to appear. The forelimbs are also developing, but they are, as yet, hidden beneath the skin.

f) The hind legs are well-formed, and the front ones appear. At the same time the lungs are forming; and, for a while, the tadpole is breathing both by lungs and gills, coming occasionally to the surface for a mouthful of air.

g) The tadpole casts its skin, and, with it, its internal gills, so that it now breathes by lungs only, obtaining all its oxygen direct from the atmosphere with the exception of some that is absorbed through its skin from the water.

h) The mouth is now wider, with well-formed jaws, and the animal is much more froglike. The tadpole has also changed its food from vegetable to animal matter.

i) The gill-clefts close up; the tail gradually disappears by absorption; the hind legs become very long and strong; and the creature, now a frog, leaves the water to search for worms, insects, etc. on land.

More stages of tadpole development and young frogs (Fig. 66 & 67)

Some of the changes noted above are to be observed only with the aid of a lens; and while the younger children will not be able to follow the whole series of events we have depicted, the senior scholars will generally do so without difficulty.

It should be noted, too, that the above summary of metamorphosis and development applies equally to the toad, there being only slight differences in appearance in the tadpoles of the toad and the frog.

When it becomes necessary to search for the spawn of the frog and the toad in ponds more or less distant from the school, one may often fail in obtaining eggs that have been very recently laid, and thus lose the chance of observing the earliest changes. This may be avoided if the school garden is surrounded by a wall or a closed fence, so that the amphibians may be kept as permanent pets in semi-captivity. In this case it is only necessary to construct a little artificial pond, which may be only a tub of water sunk into the ground, and the frogs and toads will repair it in early spring and deposit their eggs.

The changes through which young newts have to pass are also very interesting, and they present important features in which they differ from those of the frog. The eggs are of a pale yellowish color, surrounded by a gelatinous envelope, and they are laid singly on the leaves of aquatic plants. The tadpoles are of a pale color, with more elongated bodies than those of frogs and toads. The external gill-tufts persist much longer, too, even after the limbs have appeared. The front legs also appear before the hind ones, and the tail is retained in the adult as the organ of locomotion in water.

Young Newt (Fig. 70)

Since the eggs are laid singly they are not to be found without a considerable amount of searching, and they should be looked for much later than those of the other amphibians.

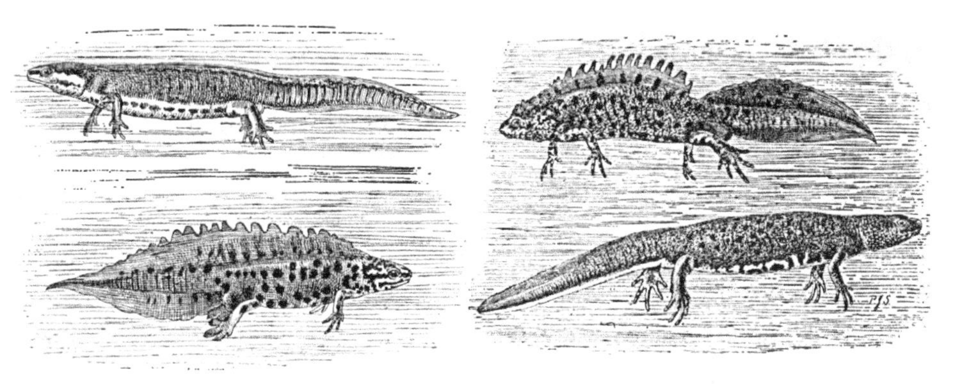

The Smooth Newt (Fig. 68) *The Great Warty Newt (Fig. 69)*

The easiest and best way to secure the eggs is to catch some adult newts with a gauze net, and put them in an aquarium properly supplied with aquatic weeds as described in Chapter 13. The deposition of the eggs may then be watched, as well as the development of the tadpoles afterwards.

VII. Reptiles

Reptiles do not, as a rule, commend themselves to teachers as objects that should be included in a school nature study course, partly, perhaps, because of a natural aversion to them, or because they are not generally well known. The British species, however, include a few creatures that are decidedly very pretty, harmless, exceedingly interesting and instructive, well adapted to a life of semi-captivity, and readily tamed.

There are not many reptiles in our country, and the complete British list runs as follows:

1. Lizards.
    a) The common or viviparous lizard.
    b) The sand lizard.
    c) The blind-worm or slow-worm.

2. Snakes.
    a) The ringed snake or grass snake.
    b) The viper or adder.

Some authorities add to the above, the smooth snake, but this reptile is very rarely seen in our country, and it is doubtful whether we should regard it as a native.

Reptiles differ from amphibians in the possession of a scaly skin; and they do not undergo any metamorphosis, the young, newly hatched from the egg, being of the same form as the parent.

Furthermore, while the lizards usually have four limbs, eyelids, and a notched (not forked) tongue, snakes have no limbs or eyelids, and the tongue is forked, with two long, slender points at the tip.

The viviparous lizard is common in most parts of the country, on heaths, moors and sunny banks; yet it is not frequently seen except by those who search for it and are acquainted with its habitats, for, when sunning itself, it is generally very inconspicuous, its colors being such that they harmonize with the surrounding herbage; and, when alarmed, it gets under cover so rapidly that there is hardly a chance of identifying it.

If it is not desired to capture a few of these lizards for close observation in a school vivarium, their interesting habits may be studied to a certain extent in the open. Having discovered one of the haunts, walk very slowly and stealthily near the sunny side of the clumps of furze or heather, or near the bank facing the sun, as the case may be, and the little reptiles may be seen basking in the sun with their bodies flattened above in order to catch as much of the warmth as possible. They will generally remain perfectly still, but sharply on the alert, while they are being watched at a distance of a few feet; but if threatened they will make a sudden dash for the neighboring cover.

They may often be caught in the hand, but they are so nimble that some dexterity is needed; and they are frequently preserved from this—the surest method of capture we can recommend—by the prickly or bushy nature of the neighboring vegetation.

The Common Lizard (Fig. 71) *The Blind-Worm (Fig. 72)*

It is only natural that those who are unacquainted with the perfectly harmless character of the lizard should endeavor to catch the creature by the tail. This, however, must not be done, for if it is, the creature will suddenly snap off that appendage and make its escape, leaving the wriggling tail in the hand. The loss of the tail is not such a serious matter to the lizard, for a new one will grow, becoming almost as long and perfect as the original one. This is a very common means by which a lizard escapes from its enemies, and thus we can account for the rather large proportion of lizards seen with a tail having dimensions something short of the perfect one.

For a closer study of the lizard it will be necessary to capture one or two and transfer them to a suitable vivarium. Instructions are given in Chapter 14 for the preparation of their home and for their management generally. They become most interesting pets, and very soon adapt themselves to their new conditions. During the first few days they will continue to be rather wild, and will not allow themselves to be handled freely. Frequently they will bite the fingers vigorously when caught, but their teeth are too small to penetrate the human skin, and therefore they are incapable of doing the least harm. In the course of a week or two, however, they become so familiar with their owner that they no longer attempt to escape when withdrawn from their case, and may then be allowed to walk freely on the hand while they are being closely examined. Their jaws are no longer used as a means of defense, they cease to snap their tail when held by that appendage, and will even take their food from the hand.

The sand lizard is very similar to the common species in general appearance and habits, but is rather larger, somewhat more snappish in its nature, and very variable in color. Some specimens are decidedly green, while others are of a brownish color, much resembling that of the last species. This lizard is to be found almost exclusively in the south of England, particularly in Hampshire. It should be looked for in sandy places, but it does not adapt itself so readily as the common lizard to a life of captivity. The eggs of this species are laid in a hole which is scooped out of the sand, and there they are left, with a slight covering of sand, to be hatched by the heat of the sun.

When searching sunny banks and hedgerows we may often meet with the pretty little blind-worm or slow-worm, basking in the sun after the manner of the other lizards, sometimes even on the top of a furze bush or a tuft of heather. It may be caught in the same manner as the legged lizards; but since it has the same habit of releasing itself by snapping off its tail when caught by that part, the precaution previously given must be observed. The blindworm comes out of its hiding place early in the morning, and makes its meal of small slugs, worms, etc.; and towards the latter part of the day, and often during the heat of the day, it retires to a hole in the ground or under the shelter of a stone. In some parts, particularly in the south, large numbers may be taken by turning over loose stones on sunny cliffs and stony banks that are not much frequented.

The blind-worm is a very interesting pet for the vivarium. It seldom attempts to bite, and when it does the bite is feeble and harmless. Like the common lizard, too, it rapidly adapts itself to a life of captivity, and will accept its food from the hand after only a few days of confinement.

We need hardly give detailed accounts of the habits and movements of the creatures we have named; for, with the hints offered above and in Chapter 14, there ought to be no considerable difficulty, as a rule, in procuring and managing them; and the rest is simply a matter of careful observation.

It is quite possible that the reader is to be counted among those who have such a natural aversion to snakes in general that they would not entertain the idea of procuring and keeping such reptiles, or of bringing them to the notice of children. But we see no reason why the study of snake life should be totally neglected. These animals possess interesting features that are entirely their own; and though one might not willingly watch their apparently cruel mode of disposing of their prey, yet their other habits are not all objectionable.

While children in town schools will find much to interest them in the movements and ways of snakes, it is really essential that country children be so well acquainted with our British species as to be able to distinguish between the harmless and the venomous.

The Grass Snake or Ringed Snake (Fig. 73)

The Viper or Adder (Fig. 74)

The grass snake is a pretty and perfectly harmless creature, and yet the majority of those who meet with it in the open consider it their duty either to run away from it or to destroy it. Even the venomous viper does not deserve the latter treatment. It will never attack a human being as long as it has a chance of escape, its venom being used only to kill its prey and to protect itself when hard pressed.

Both species are very common in many parts, and therefore all country children (and town children who visit the country) should be so familiar with them that they can identify them at a glance.

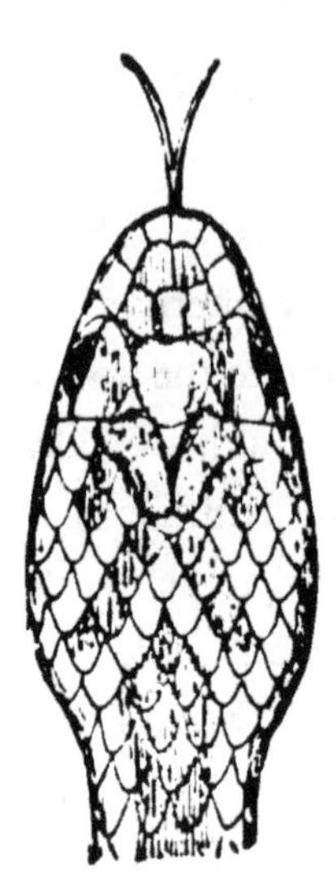

Viper head (Fig. 75)

The grass snake, which grows to a length of three feet or a a little more, may be seen sunning itself in spots that are not far removed from water, and often in the water itself, where it swims with a waving motion of the body. It may be distinguished from the viper by the yellow patch just behind the head, and by the absence of black markings down the middle of the back. When captured it will hiss and also eject an evil-smelling fluid as a means of defense; but it never bites, and its teeth are too small to penetrate the skin if it did.

The viper is a smaller snake, seldom growing much over two feet in length. It has a black V-shaped mark on the head, a conspicuous, zigzag, black line down the middle to the back, and a short, tapering tail.

Its weapon of defense and offense is not a sting, but a pair of long teeth or fangs in the upper jaw. These fangs are on movable joints, are depressed against the jaw when not in use, and erected when required for action.

When an adder is about to attack, it opens its mouth to a very wide angle, and then strikes (not bites) its victim with its upper jaw, driving the fangs into its flesh. At the base of each fang is a poison gland, and the venom produced by the latter passes down a little canal in the fang, and then into the wound.

We do not recommend that a live viper be kept in a vivarium for the observation of children unless the teacher who introduces it is thoroughly familiar with snakes, and keeps his specimen securely locked up; but there is no reason, it seems to us, why the grass snake should not be so kept. It soon becomes very tame, so much so that it will allow itself to be handled freely without making any attempts to escape.

All necessary instructions are given in our chapter dealing with the vivarium, and the creature's habits should be made a matter of direct observation as before suggested. In order that its means of locomotion may be understood, however, we will call attention to the very large, overlapping scales on the undersurface of the body. These have their free edges directed backwards, and each one is attached to a pair of ribs by which it can be moved. Thus the snake pushes itself along the ground or among the herbage by means of the edges of the lower scales, but it will be observed that the creature is greatly assisted in its progress, whether along the ground or climbing among the herbage and bushes, by the manner in which it curves its body.

Where the school garden is enclosed by a moderately high wall, or a closed fence that admits of no outlets, lizards and harmless snakes may be kept in a condition of semi-captivity. This is undoubtedly the best way of keeping such creatures under observation, for, the conditions under which they live being an approximation of those which obtain in Nature, their habits will also be approximately normal; and the children are sure to take great interest in searching out the animals from time to time, and watching their movements.

If this plan is to be carried out, build up a loose rockery of piled stones in a sunny corner, surround it with a liberal growth of small perennial shrubs, such as furze, bramble and, if possible, heather, and allow the grass and other weeds to grow as they will. The corner itself should not be watered other than by the rain, but sink a shallow pan or other vessel into the ground to provide a little water for drinking and bathing.

It should be remembered, however, that lizards are good climbers, and can easily scale an ordinary wall; also that snakes will readily escape if a tall herbage or shrubby growth is present to assist them in their upward progress. To prevent such escape of the captives, allow no vegetable growth to reach within several inches of the top of the wall, and place on the latter a coping of slates or other material that projects a few inches on the inside.

Frogs, toads, newts, and various other creatures may be similarly kept in the school garden, but the amphibians and other animals that require protection from the direct rays of the sun should have their rockery and surrounding herbage in a shady corner. Of course we can hardly hope to give here ample instructions for the care of all creatures that may be chosen for such a life of semi-captivity, but there will be, as a rule, no difficulty in ensuring success if we make it a point to follow Nature—to keep each animal under conditions approximating, as near as possible, to those under which we find it in its wild state.

VIII. Birds

Leaving, for the present, the study of the general structure and habits of birds, we will deal briefly here only with certain occupations connected with bird life that belong especially to the spring season. The many species of birds that find a place in the British list may be roughly classified into (1) the residents, which remain with us throughout the year; (2) the summer visitors, that leave us for warmer climes in the autumn, and return to our country in the following spring; (3) the winter visitors, which leave us for the North in the spring, returning when the cold weather again sets in; and (4) the casual visitors which do not make Britain their permanent home, though they may more or less frequently pay us a visit and even, at times, rear their broods here.

The Song-Thrush (Fig. 76)

Many of our resident birds may be studied with advantage in winter and early spring, for at this season they have often considerable difficulty in obtaining the necessary amount of food especially during prolonged frosts, and so are constrained to approach the dwellings of man both for food and shelter.

Such birds may be encouraged to frequent the neighborhood of our houses by placing one or more feeding tables in suitable positions,

The Robin (Fig. 77)

The Great Tit (Fig. 78)

and keeping the same supplied with food adapted to the requirements of the birds we hope to entice, together with a shallow vessel of clean water.

Very good feeding tables may be constructed by fixing one end of a broom-handle in a hole bored in a wood slab measuring about a foot square, and shaving off the other end to a point for driving into the ground.

The finches and other hard-billed birds may be attracted to the table by means of a supply of breadcrumbs and mixed birdseed, while the insectivorous species will show their appreciation of chopped meat, or of a bone with a little meat on it. The pretty little blue-tit will pay frequent visits to a piece of fat meat or bacon rind suspended by a string, or half a coconut similarly attached to the branch of a tree.

A feeding table for the birds (Fig. 79)

In short, having ascertained, by experiment, or from other sources, the nature of the food of birds that frequent the neighborhood of the school, much may be done during the winter and early spring to encourage them to come within the range of easy observation.

The Wren (Fig. 80)

The Swallow (Fig. 81)

A little later in the season, when the summer visitors begin to arrive, and when preparations carried on, especially by the children of schools in the country and on the outskirts of towns.

Children should be encouraged to watch the birds closely at this season. They should endeavor to identify those which are seen sufficiently near for this purpose—to know them by their size, form and color, by their mode of progression on the ground, by the peculiarities of their flight, and by their song.

The House Martin (Fig. 82)

The Skylark (Fig. 83)

The arrival of the summer visitors will be noticed, and the dates of the first appearances entered in the nature diary. Similar entries will also be made of the observations relating to the nesting, including notes on the materials collected by the various species for the construction of their nests, and the manner in which these materials are secured.

In many cases it will be easy to follow the flight of the old birds, and so to become acquainted with the site of the new home. The building of the nest may then be watched day by day, or as opportunities arrive. The building over, note may be taken of the dates on which the eggs are deposited, the date when the young birds appear, and the whole progress of the young till the time when they entirely discard their temporary home. During this season the movements of the parent birds are also peculiarly interesting. The part taken by both male and female parents in the charge of the nest and in the feeding of the young ones should be noted, and also the interesting movements of the parents as they teach their family to shift for themselves.

The Barn Owl (Fig. 84)

The examination of bird nests during the breeding should be conducted with great care, for many birds will entirely forsake the nest if frequently disturbed, more especially if the nest itself and the surroundings show signs of interference. The few illustrations interspersed in this section will enable the reader to identify some of the common birds that frequent the neighborhood of our dwellings.

IX. Mammals

Mammals may be studied almost continuously throughout the year, but during the spring we commence, or resume, our observations of the wild species that have been but little seen during the winter months, partly on account of their hibernating habits, and partly because our outdoor studies have been greatly retarded by uncongenial weather.

It is common to confine the school studies of mammals exclusively to the domestic species and others that are kept in a state of captivity. This, we think, is quite a mistake, at least as regards those schools which are within easy distance of fields and woods.

Children should be encouraged to search for some of the few small wild mammals we possess, in their haunts; and, having discovered them, to remain quite still for a time, in some place of concealment, watch their movements, and endeavor to trace them to their dwellings, thus discovering the nature of their homes and the manner in which they bring up their young ones. In this way children may learn many interesting things concerning our wild creatures that could not be learned in any other way.

Several of our wild mammals produce their young in spring and early summer, and during this nursing season they may be more easily watched than at any other time, for they will never stray far from their homes as long as they have their young to protect and feed. The old hedgehog may be followed to her nest containing a number of little ones as yet with a smooth and spineless skin; and if observed from day to day she will become so familiar with the monster who watches her that, after a time, she takes but little notice of the intrusion to her beat and her nest. Squirrels may also be traced to their nests in the trees, and watched as they repair to their winter stores of food not yet exhausted; and the frolics of the young ones may be observed as they chase one another among the branches. These and many other similar observations belong essentially to spring and early summer.

Special attention should be called to the play of young mammals, and the children should note that the movements made by the young of each species in their frolics are exactly those on the skilled performance of which the mature animal owes its very existence. Thus, the young lamb skips, training itself for the jumping that becomes necessary, at least in the wild state, for the tufts of scanty herbage that grow on mountains, precipitous hills and cliffs. The play movements of the kitten also provide the precise training required by the cat in the capture and disposal of its prey.

In Chapter 16 we give some practical hints relating to the management of those species which it is desired to keep in captivity for more detailed and continuous study.

D. The Sky and the Weather

Many of the features of the sky and the weather are of such a general nature that they may be studied almost equally well throughout the year. Of such are the changes of the moon, the observation of the planets as they become visible, the apparent movements of the heavenly bodies due to the rotation and revolution of the earth, including the division of the visible stars into two groups—those which rise and set, and those which never set; also the study of winds and clouds, sunshine and rain, etc.

During the spring, however, special attention should be drawn to the apparent variations in the path of the sun. Let the children make all their observations from the school or from some other suitable fixed position, and note the point on the horizon, located with the aid of some more or less conspicuous landmark, at which the sun rises, the height of the sun at midday, and the point on the horizon where he sets. These observations should be entered in the nature notebooks, and compared with the results of similar observations made on other days, at fairly regular intervals, as the spring advances. Such observations will, of course, be more or less interrupted by misty or cloudy days, but sufficient opportunities will occur to provide all that is required to give a general idea of the apparent movements of the sun, to work out the causes, and to study the effects. Let them see clearly that, other conditions being the same, the warmth of the sun increases with the altitude of that body in the sky.

Even the youngest children may take part in observations of this kind, and note how we are affected, in a general way, by them; while the senior classes may, with the aid of a globe or a ball to represent the earth, and a distant, fixed point to denote the position of the sun, be led to work out the causes of the apparent movements observed.

Simple arrangement to find planet altitude (Fig. 85)

The elder children, too, should be taught some simple method of measuring, approximately, the altitude of the sun or other heavenly body. Thus, a very narrow paper tube may be fastened in such a manner that it can be made to revolve on a point at the edge of a level table or bench. This tube, used after the manner of a telescope, is placed in such a position that the body in question may be seen on looking through it, and the angle between it and the level table-top, representing the altitude of the body under observation, may then be read off on a large protractor. We need hardly mention that when such an observation is being made of the sun itself, the eyes should be protected by means of a piece of deep blue or smoked glass.

Again, while even the young children are made to observe that the lengths of shadows decrease as the altitude of the sun increases, the senior scholars will be taught to make fairly accurate measurements of the altitude from the length of the shadow of an upright stick, such shadow being made to fall on a level surface.

Furthermore, while younger children may be able to grasp some elementary notions concerning the nature of the stars—that they are immense, intensely-heated globes like our sun, at such enormous distances from us that they appear as mere points of light; that they are, in fact, distant suns, probably, for the most part, like our sun, the centers of other systems of worlds resembling the system formed by our sun and the earth and other planets revolving around it—the elder ones may be taught to recognize some of the principal constellations or groups of stars, to call them by their names, to note how they all revolve around the pole star as a center, and to observe that while some of them, in the course of their daily revolutions, always remain above the horizon, so that they are visible on all clear nights throughout the year, others revolve in larger circles, rising and setting like the sun itself.

Then, as the season advances, the elder scholars will note that these rising and setting constellations appear to change their positions a little, night by night, on account of the annual motion of the earth around the sun; that some, visible just above the horizon during early spring, cease to appear as time goes on, and remain beyond our view until the end of the year or the beginning of the following year, while new constellations make their appearance as the former disappear. The senior scholars will then attempt to work out the explanations of these apparent movements of the fixed stars, under the guidance of, and, as necessary, with the assistance of, the teacher; and thus they will be led up to the uses of the astronomical globe, star maps, and the stellar planisphere.

Of course we are aware that many teachers, admirably qualified for their work in a general way, are not sufficiently acquainted with the elementary principles of astronomy to deal confidently with some of the matters to which we have referred as suitable for the investigation of senior scholars; but the teacher who is interested in Nature, and loves to help in the expounding of her works, will find it no hardship to devote a small amount of time to the study of certain branches of the subject in order that he may be better able to open up the minds of the children under his care.

As to the study of the weather, there is little peculiar to the spring season beyond the gradual increase in temperature due to the increasing altitude of the sun and the gradual lengthening of the days, interrupted or modified at times by other factors that help to determine atmospheric conditions; so that the general remarks made on the study of the weather apply equally to this and the other seasons.

Attention will be called, however, to the characteristics of the spring showers, and to the rainbows which so frequently accompany them. As regards the latter, although almost everybody (we have met with many children living in densely-populated cities and towns who have never seen a rainbow) is familiar with the appearance of the rainbow, comparatively few persons have noted the exact conditions under which it becomes visible to us.

Diagram illustrating the conditions under which the rainbow is formed (Fig. 86)

S, rays from the sun; *p* & *n*, raindrops; *O*, the observer. The outer and larger is the fainter, secondary bow.

Hence, the following points should be worked out as far as they fall within the capacities of the children:

1. A rainbow is produced whenever the direct rays of the sun fall on raindrops.

2. It is seen under the following conditions: We stand with our backs to the sun, and rain is falling in front of us, but not necessarily on the spot where we stand.
3. If one bow only is visible, the colors are always in the same order, with the red outside and the violet inside.
4. When, as occasionally happens, we see two bows, one within the other, the outer or secondary one is much fainter, is made up of the same colors, but the order is reversed.
5. That the rainbow encloses the base of a cone of which the eye of the observer is the apex; and the axis of the cone is parallel to the sun's rays. Hence, the lower the sun, the higher and larger is the bow.

These conditions may be verified on any bright day by the formation of an artificial rainbow in the school playground. This is easily done, where the water-pressure is sufficient, by sending a fine spray from the garden hose high into the air.

The senior children may also be shown simple experiments illustrating the decomposition of light by means of a glass prism, and the reflection of light by means of a mirror; and thus they may be led to understand how that the light, decomposed as it enters the raindrops, is reflected back to them from the inside faces of these drops.

—5—

Summer Studies

A. General Remarks

Summer is the season of the greatest profusion of both animal and vegetable life. The accumulated energy resulting from the great length of the days and the great altitude of the sun not only produces an abundance of living things, but also gives rise to the general activity and rapid development which characterizes this period of the year.

No difficulty will now be experienced in the collection of suitable material for nature lessons given in the school; and the profusion of living things, both of the animal and vegetable worlds, is so great that the teacher will often find it necessary, during the progress of outdoor work, to guard against indiscriminate collecting and aimless observations.

B. Plant Life

I. Wild and garden flowers

At this season of the year, when so many plants are in their mature condition and flowers are to be found in abundance, much time may be spent in studying the general build of plants and their habitats and habits.

The greater part of this work should be carried on, if possible, outdoors, either during the rambles of the children or in the school garden; but when the weather or other circumstances preclude outdoor studies, whole plants may be examined within the school with the object of noting the various parts and their arrangement.

The study of plants thus isolated from their natural surroundings is, however, far less instructive and much less interesting than the study of the same species in their natural habitats, and every available opportunity should be secured for the observation of plants under their natural conditions, even if there is no better field than a patch of waste ground, a weedy wayside, or the neglected corner of a garden.

The Common Mallow (Fig. 87)

The Bird's-Foot Trefoil (Fig. 88)

Although we do not wish to discourage the careful cultivation of garden flowers, especially with regard to those localities where horticulture is one of the supporting industries of the inhabitants, we feel bound to state that the child will get a far truer idea of the interesting habits of plants by the thoughtful observation of a wild patch of ground than by the study of many of the florist's productions. In fact, many of our valued garden flowers are so altered by the florist's art that their original characteristics are to a great extent destroyed.

Thus, the standard rose, although an object of great beauty, is a more monstrosity, retaining but little of its original form and habit, with no power of reproducing its species; while the common dogrose of our hedgerows is rendered very interesting by its climbing habit as it struggles upward among the surrounding shrubbage for its share of light and air, and its beautiful flowers are perfect in all their parts, giving rise to interesting clusters of fruits in due season.

The Silver Weed (Fig. 89)

The Cow Parsnip (Fig. 90)

Again, among the wildflowers we are able to witness the many interesting ways in which the plants have to struggle for their existence, to see how the battle of life terminates in the survival of the fittest, and to note the ways in which various species protect themselves from their foes; while in the well-kept garden the conditions of growth are so altered that the above features are to a great extent lost. Yet there is certainly much to be learned from the ordinary garden of cultivated flowers, shrubs and trees.

If the children are allowed to do the whole work of the garden themselves—if they are taught to study the soil, to prepare it for the flowers or crops, to watch the effect of the application of both organic and mineral manures and fertilizers, to raise plants from their seeds, to attend to the thinning and transplanting, to watch the effects of the pruning of branches and roots, etc., they will learn a great deal concerning the nature of plants and of the conditions that affect their growth and productiveness, and will, at the same time, gain much knowledge that will be useful to them in the future.

As we previously hinted, there is much to be learned by the observation of wild Nature during school rambles that cannot be learned in the cultivated and well-kept garden, and we shall now deal with some of the more prominent features concerning plants to which the particular attention of the children may be directed.

The Teasel (Fig. 91)

The Burdock (Fig. 92)

First, then, as regards the habitats of plants, we should be careful to see that the soil and the situation of each species are noted. Some plants will grow in almost any kind of soil; others are very partial to one particular class of soil but are not restricted to it; and not a few are to be found always on the same kind of ground and are therefore more local in their distribution. Such observations should be entered in the children's notebooks, and the scholars should also be encouraged to classify the species observed according to the situations in which they generally occur, associating them respectively with the meadow, the riverbank, the marsh, the wayside, the shady hedgerow, the sunny bank, the wood, the moor, etc.

The senior children may also be taught to note how certain plants are peculiarly adapted to the situations in which they occur, and to observe how the same species will vary in structure and habit when it grows in different situations. Thus, they may be led to see that certain plants which grow on walls, roofs, cliffs, and other very dry places, produce very thick, fleshy leaves which store much moisture to enable them to live through periods of drought;

and that when these same species are seen in moist situations where the necessity of storing moisture no longer obtains, the leaves become thinner and less succulent.

As another example we may mention the common rest-harrow, which, in dry situations, has many very short branches, terminating in a thorn, and also very small leaves, so that but little surface is exposed for the transpiration of moisture; while in moist situations the same plant produces well-developed branches and larger leaves.

The Ragwort (Fig. 93)

The Scentless Mayweed (Fig. 94)

The Milfoil (Fig. 95)

As a part of this same study the children will note that certain different species of plants are frequently to be found growing together, and they will endeavor to determine whether, in the case of two or more species thus associated, the association is advantageous to either or both. It is probable, for example, that the white dead-nettle, so often seen accompanying the stinging-nettle, is to a great extent protected by its resemblance (as far as the leaves are concerned) to its virulent companion.

The study of plant associations is a very engaging one, and the teacher will do well to direct special attention to it, for it presents to us many interesting problems, solved and unsolved, all of which afford much food for thought.

Let the children distinguish between the erect, decumbent, prostrate, trailing, and climbing plants, noting, in each case, how the plant manages to secure the necessary supply of light and air. Runners and prostrate and trailing stems should be traced with the object of noting to what extent extra roots are produced for the absorption of moisture from the soil.

The Yellow Toadflax (Fig. 96)

The Kidney Vetch (Fig. 97)

The Melilot (Fig. 98)

Climbing plants in particular will afford a vast amount of useful study, especially if their habits are watched at frequent intervals from their seedling stages, at which time they are generally quite erect, and show no trace of their future climbing habits. We cannot attempt to give here a descriptive account of the many ways in which the weak-stemmed plants support themselves, but we give an outline of the principal observations that might be made with regard to them.

1. Twining stems, as in the hop, convolvulus, honeysuckle, black bindweed, etc. The direction of the spiral—right-handed or left-handed. The formation of ropes for mutual support when no other support is near. The revolution of the tip of the stem when seeking an object on which to climb. The behavior of the plant when it fails to find such an object. Interesting experiments might be made with the view of determining the size-limit (diameter) of the prop around which the stem can form its spiral, and also to test the limit of the inclination of the axis of the spiral in any one species. (Some will climb around horizontal supports; others only along props that do not deviate beyond a certain angle from the vertical line.) (Figs. 100, 101, and 102.)
2. Tendrils, as in peas and vetches. The direction taken by the tendrils. Their sensitiveness to touch. Note whether the tendrils are modified stems, or leaves, or flower-stalks, or leaflets of compound leaves.

3. Twining leafstalk, as in the wild clematis. (Fig. 103.)
4. Clinging suckers—the ivy.
5. Supporting prickles (outgrowths of the epidermis) curved downwards to form hooks, as in the dogrose. (Fig. 104.)
6. Supporting bristles on stems or leaves, or both, as in the goosegrass. (Fig. 105.)

In all the above cases the children should discover, in the slender stems, the necessity of the climbing habit in enabling the plants concerned to compete with their neighbors in the fight for light and air.

Bindweed stem, twining left (Fig. 100) *Hop stem, twining right (Fig. 101)* *Wild Rose prickles (Fig. 104)*

The manner in which many plants protect themselves from their herbivorous and other enemies will form another useful study—the thorns and prickles which prevent certain species from being trodden down or devoured, the hairs which prevent the passage of slugs and snails that would eat the leaves or flowers, and the objectionable tastes and odors and the formation of poisonous products which also serve to prevent destruction by herbivorous creatures. The elder children might also be taught to observe that thorns are modified branches—branches that are produced from buds, but which, though they sometimes give rise to other buds and leaves, gradually taper off to a sharp point; that spines are projections from leaves, or even complete leaves from which the blade has disappeared; that prickles are mere outgrowths of the epidermis, with no fibers continuous with those of the structure on which they grow, but are easily detached, leaving a clean scar; and that hairs and bristles are modified cells of the epidermis.

The forest trees and shrubs being now all in full leaf, and herbaceous vegetation in its greatest profusion, this season is decidedly the best for the study of the forms of leaves, and of the different ways in which the leaves are disposed.

The Black Knapweed

(Fig. 99)

Hop stems in twining position

(Fig. 102)

The Wild Olematis

(Fig. 103)

As regards their forms, the difference between simple and compound leaves (Figs. 107 and 108) should be pointed out, and all the gradations between the two, from the entire or slightly cut margin to the completely divided blade. The self-clinging Virginian creeper, so largely grown in our towns, affords a splendid example of the transition from the simple to the compound leaf, for both, together with intermediate stages, are all to be found on the same tree. Attention will also be called to the different kinds of surface exhibited by leaves, including their clothings of bristles, hairs, down, etc. and attempts will be made to solve the problems connected with the uses of such clothing to the plants concerned.

The study of the arrangement of leaves is even more instructive and, considered in conjunction with the fact that the leaves are the factories in which the various compounds required by the plant are built up under the influence of sunlight, this becomes a most important and interesting matter for close and thoughtful observation.

Here, again, it is impossible to give a detailed account of:

1. In many herbaceous plants the lowest leaves are largest, or have longer stalks, so that they are not shaded by the leaves above them. The upper leaves are also directed upwards, while the lower are horizontally disposed, and thus the former do not overshadow the latter.
2. When the leaves of a plant are mainly or entirely close to the ground, they are usually arranged in the form of a rosette, with little or no overlapping. (Fig. 109.)
3. Where the top leaves are rather densely clustered, they are also arranged in a rosette, without much overlapping. (Fig. 110.)
4. The leaves that are supported on an erect stalk, if not much divided or cut into, are usually disposed along spiral curves, by which arrangement they receive the maximum of light. If, however, the leaves are much divided, so that light can pass freely between their lobes, there is no necessity for such an arrangement.
5. In very leafy trees, such as the beech, the upper branches are directed upwards, the next obliquely, and the lowest horizontally—another arrangement by which the leaves secure a maximum of light.
6. The arrangement of leaves on vertical stalks is different from that on horizontal stalks of the same plant or tree, the difference being again brought about in order to expose the greatest possible leaf-surface to the light.
7. Interesting leaf-mosaics should always receive attention. In the elm and the beech, for example, we note leaves of different sizes, so disposed that one does not overshadow another, and leaves are often turned backwards away from the tip of the twig, in order to fill in spaces left between other leaves, rather than to lie above or beneath them. Some of the most striking examples of leaf-mosaics are to be found in woods and other shady places, where every effort has to be made to secure sufficient energy from the sun. (Fig. 113.)

Under the influence of the sun's light and heat, the summer leaves are rapidly building up various organic compounds from the simpler mineral compounds absorbed from the soil and the carbonic acid gas taken in from the atmosphere. In many cases the material manufactured is more than is required for the immediate needs of the plant, and consequently is stored for future consumption. The observations of children should be kept alive to this matter of storage, especially as regards biennial and perennial plants; and they should be directed to note the stores laid up in the thickened rootstocks, tubers, corms, rhizomes, bulbs, etc., during the summer months.

The Goosegrass (Fig. 105) *The White Bryony (Fig. 106)* *Small-flowered Crane's-bill (Fig. 109)*

We now come to the consideration of the flowers, the beautiful forms and colors of which are so attractive that they are frequently, but wrongly, studied without regard to the nature of the plants to which they belong, and equally regardless of their habitats.

The parts of the flower, their arrangement, relative positions and uses should be studied with due regard to the age and capacity of the children; and since the functions of the various parts are not always to be understood without a careful observation of the development of the flower, the gradual unfolding and maturing of the latter should be carefully watched either outdoors or by the observation of plants that are grown in the school.

Technical terms need never be introduced. With the younger children they should never be used; but rather encourage the children to describe what they see in their own simple language. Let them give their own names to the parts described, correcting them only when such names are inappropriate. Adopting this plan, the teacher will often find that the younger scholars will hit upon simple names even more appropriate than those he himself would suggest.

The younger children may be made to understand something of the use of the pollen cells and of the ovules within the ovary—the necessity of the former in causing the development of the latter, and simple experiments (such as the removal of the anthers before the pollen cells are set free, together with the exclusion of insects by means of gauze coverings) may be made to demonstrate the function of the pollen; but the more detailed account of pollination and fertilization should be left for the consideration of the senior classes.

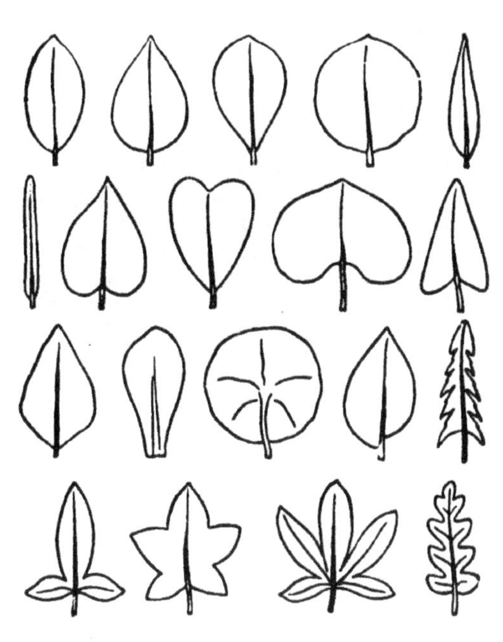

Various forms of simple leaves (Fig. 107)

During the study of flowers outdoors, all the children should be encouraged to watch bees and other insects at work; but while the little ones will get only very rudimentary ideas concerning the work of the insects in the transmission of pollen from flower to flower, the elder children will note how some flowers are so constructed that certain species of insects only can effectually work for the good of the flowers in question, that many flowers are specially provided with the means of attracting the very species which work for their good, and that the relation between the flowers and the insects attracted by them is usually one of mutual benefit.

Compound leaves (Fig. 108)

This subject of the relationship between flowers and insects is so full of interest, and presents so many varied aspects, that the teacher who is not well acquainted with it will do well to study the matter from one or more of the many good books dealing with this topic, in order that he may be better prepared to answer the numerous questions that are likely to arise from enthusiastic and thoughtful scholars.

Leaves of Sycamore branch tip (Fig. 110)

The distribution of pollen by means of the wind is another matter for careful study. The children should be led to discover the features by which we can distinguish between the flowers that are aided by insects and those which are not. They will note, in the latter, the absence of gaudy colors, the absence of nectar, and the presence of an abundant supply of pollen to compensate for the waste that must necessarily accompany wind distribution.

The Great Knapweed (Fig. 111)

The Self-Heal (Fig. 112)

A twig of the Hazel (Fig. 113)

During the examination of flowers the younger children should never be allowed to pull their specimens to pieces; and the dissection of flowers is seldom necessary with the elder children unless they are sufficiently advanced to make a study of the relationships of plants as far as they are revealed in the number, form, structure, and arrangement of the floral organs, in which instance the necessary dissections should be carried out in a systematic manner with every attention to accuracy and neatness, so that the work performed may be of additional value as a training towards acquiring manipulative skill.

The Harebell (Fig. 114)

The Tormentil (Fig. 115)

The Woundwort (Fig. 116)

Finally, as regards the outdoor study of flowers, whether pursued during school hours under the supervision of the teacher or in the children's own spare time, the notebook should always be used for the purpose of recording the chief points of interest connected with the species seen, and of making sketches to illustrate the more interesting features of their structure. For this purpose perhaps the best kind of book is one in which leaves of ruled writing paper and drawing paper alternate. The former may be divided into columns headed as follows, and the sketches will be made on the opposite page of drawing paper:

1. Date
2. Name of flower
3. Where found
4. Soil
5. Remarks

The illustrations accompanying this chapter will enable the reader to identify some few of our commonest summer wildflowers.

II. Forest trees and shrubs

All our trees and shrubs are now in full leaf. Some have long since produced their flowers, a few are in flower, and in the majority of them the fruits are reaching maturity.

The Meadow Sweet (Fig. 117)

The Wild Carrot (Fig. 118)

The Woody Nightshade (Fig. 119)

The following is a summary of the principal occupations during the summer rambles.

1. Observe the trees from a distance in order to study their general form and appearance.
2. On a nearer approach, observe the nature of the bark, the mode of branching, the direction of the higher and the lower branches, and the forms of the leaves.
3. Study the nature of the developing fruit in cases where such is to be found.
4. Note the positions of the buds—terminal and axillary—which are destined to produce new branches in the following year.
5. The senior children will learn to distinguish between true fruits and those structures which, although apparently fruits or parts of fruits, are really developments of parts of the flower other than the ovary. They will note, for instance, that the prickly mast surrounding the beech fruits (nuts), the woody scales of the fruiting catkins of the alder, the leafy structures behind the fruits of the hornbeam, and the cupules of acorns are not parts of the respective fruits, but are modified scales or bracts of the flowers.

When passing a woodman at work among the trees, or the sawyer engaged in his pit, the children will make use of the opportunity of studying the wood of the felled trees, and particularly of noting the transverse sections of the trunks and branches which teach so much concerning the growth of the trees.

III. Plant activities

We suggested, as suitable occupations for the spring season, simple experiments for the investigation of the absorption of water by roots, the transpiration of water by leaves, and the direction of the flow of the sap in trees. These occupations may be repeated in the summer if such repetition is considered advisable, or may be performed during the summer months if they were omitted in the spring.

To these may be added other experiments demonstrating the formation of organic products in the leaves, the necessity of light in connection with these functions, and the storage of the products in roots, tubers, etc. Among such experiments the following may be chosen:

The Water Figwort (Fig. 120)

The Sow Thistle (Fig. 122)

The Small Bindweed (Fig. 121)

1. Remove a leaf from a growing plant towards the end of the day, the plant having been exposed to sunlight during the whole of that day. Kill the leaf by immersing it in boiling water for one or two seconds only, and then place it in methylated spirit, keeping it in this liquid until all the green coloring matter (chlorophyll) has been dissolved out, and the leaf has assumed a very pale color. Now put the leaf in a solution of iodine, made by dissolving a very little iodine in a weak solution of potassium iodide until the mixture is only of a pale sherry color. The leaf now darkens to a bluish color, proving the presence of starch.
2. Cover the same plant with a box, to exclude all light, for about twenty-four hours, or place it in a dark cupboard for the same purpose, and, at the end of that time, remove another leaf for exactly the same treatment as the first. This second leaf does not turn to a bluish color when treated with the iodine solution, proving that starch is absent. We thus prove that starch, formed only during the exposure of leaves to light, is absorbed by the plant during intervals of darkness. (The results of these experiments are rendered still more striking if, instead of using two separate leaves, we shut out the light from a portion of a leaf for about twenty-four hours, by covering that portion with brown paper or some other opaque substance. After this leaf has been killed, decolorized, and treated with the iodine solution, the part previously shaded will reveal the absence of starch, which will be proved to exist in the remaining portion that was exposed to light) Some leaves give better results than others in these experiments, and it will therefore be interesting to select leaves from different plants and to compare the results obtained.
3. Cut a thin slice of potato, and pour upon it a few drops of the iodine solution. The dark blue color produced proves the presence of stored starch.
4. Apply the iodine test also to the fleshy cotyledons of a fresh bean, or of a bean that has been soaked in water for several hours. The result reveals a store of starch reserved for the early growth of the young bean plant.

The Spear Thistle (Fig. 123)

The Tansy (Fig. 124)

IV. Flowerless plants

We have hitherto referred only to flowering plants and trees, but there is no reason why the children, and especially the elder ones, should not have some rudimentary acquaintance with common flowerless species such as ferns, mosses, fungi, and algae. These plants are produced from spores which, unlike seeds, are minute simple cells that do not consist of or contain an embryo plant; and their interesting life histories must prove very instructive to children.

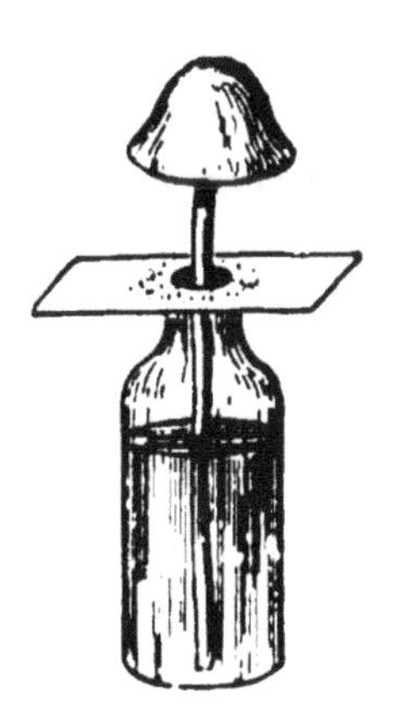

Tool for collecting spores from a fungus (Fig. 125)

Let the children examine a moss plant, and obtain the spores by shaking ripe capsules over a sheet of white paper.

A typical fungus, such as the mushroom, may also be observed, and the spores may be obtained from it by placing its stem in a bottle of water, the mouth of the bottle being covered with a small piece of white paper through a hole in which the stem passes. This arrangement is shown in the accompanying figure. The spores when ripe, fall on the paper from the gills of the underside of the disc (*pileus*).

The Dogwood in flower (Fig. 126)

The Broom (Fig. 130)

The life histories of flowerless plants are sometimes not easily observed and, therefore, fall somewhat outside the work of school nature study; but we may certainly make an exception in the case of ferns, for these are easily reared from their spores, and their development may be watched without the aid of the microscope.

An arrangement for the propagation of ferns from the spores (Fig. 128)

The spores of ferns may be obtained in abundance by putting some fronds with ripe spore-clusters (*sori*) in a large paper bag, allowing them to dry thoroughly, and then, after tying up the mouth of the bag, giving the whole a thorough shaking. After allowing the bag to remain perfectly still for some time, a quantity of the spores, in the form of a fine dust, will be found at the bottom.

Now arrange the apparatus shown in the accompanying illustration, consisting of a shallow tray or dish containing a little water, a piece of unglazed brick or tile, and a bell jar; or, if the tray or dish is a moderately deep one, a plate of glass may be used instead of the bell jar. Sprinkle water on the top of the brick or tile (which must not be submerged), scatter a little of the spore-dust on it, and cover the whole with the bell-jar or glass plate.

The Elder (Fig. 127)

The Guelder Rose (Fig. 129)

The spores will soon begin to grow, giving rise not to ferns, but to little leaf-like bodies (*prothalli*) on which are male and female organs. After the latter have been fertilized by the former, a little fern frond will commence to grow from each prothallus, and the young ferns may then be transferred carefully to a rich, peaty soil for further development.

The arrangement described above may be modified by using peaty soil in the place of the brick or tile, in which case the ferns may be left to grow to a later stage before transplanting; or, with judicious thinning, may be allowed to remain permanently on the original ground. And, if soil is used, it should be well pressed to a smooth and level surface before sprinkling with the spores.

It will have been noticed, from the foregoing description, that a fern spore does not give rise at once to a fern, but to a minute leaf-like prothallus from which the young fern afterwards sprouts. This is an interesting example of the 'alternation of generations' that is common among the flowerless plants.

C. Animal Life

Animals of all kinds are also exceedingly abundant during the summer months, and animal activities are now at their maximum. Many species, indeed, are so plentiful, in the immediate neighborhood of our dwellings as well as in the open country, that it will be well to give our attention almost entirely to the wild creatures, leaving, for the present, the study of domestic animals which can be pursued almost equally well throughout the year.

Again we would remind the reader that we do not attempt the impossible task of including, in the space at our disposal, information on all the common creatures which may come within the children's observations; our object is to give the assistance and advice necessary to enable the teacher properly to direct the children in their observations and investigations. Furthermore, it is to be noted that the teacher should observe as well as the children, for no student of Nature is so thoroughly versed in the peculiarities and habits of living things that he can afford to stand by while his class is at work. The ideal nature lesson is one in which both teacher and children work together, observing and discussing. The teacher does not simply prepare a lesson and give it. The lesson develops as it proceeds, the teacher, of course, making abundant use of his previous knowledge obtained by former observations of the things concerned, as well as by his book lore.

The general hints given for the spring studies of animal life apply equally well to the summer, but some matters previously alluded to may be emphasized here, and others will be introduced according to the requirements of the season.

The small creatures of the garden will now supply an abundant source of interest; and, of course, a still greater variety of species will be found during rambles beyond the confines of the town.

Almost every overturned stone or clod of earth will reveal some living thing of interest. Under projecting ledges and in various sheltered places we find numerous nocturnal creatures taking their daily rest; and these are frequently so unwary that they may be approached closely enough to be examined through a magnifying lens.

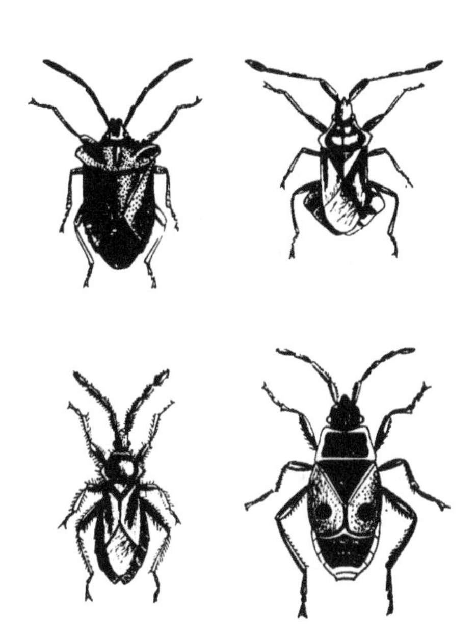

A group of plant bugs (Fig. 131)

On the foliage of plants and trees are many species of caterpillars, some feeding during the day, others, of nocturnal habits, remaining at rest as long as the day is bright. These, either watched at frequent intervals outdoors, or, perhaps, with greater convenience, kept under observation within the school building, will exhibit remarkable metamorphosis, finally assuming the form of a moth or a butterfly.

In similar situations we observe some species of plant-bugs, some of them really beautiful insects, which thrust their sharp beaks into the stems and leaves of plants and suck out the sap; also the larva of the frog-hopper or cuckoo-spit, which protects itself by ejecting a fluid in such a manner as to produce a series of little bubbles that eventually completely cover its body, while, thus hidden from view, it sucks the sap from the plant on which it is situated. A continuous observation of one of these creatures will show that it changes to a winged insect with prodigious leaping powers.

Furthermore, various species of aphides or plant-lice may be seen in crowds, each individual with its beak thrust into a tender plant or twig, busily extracting the sap to the great annoyance of the lover of flowers; and, among these, we often meet with the larva of the lady-bird—a pretty little beetle, the bitter enemy of the aphides, which it greedily devours, and therefore a friend of the florist and the gardener.

Among the twigs and stems of the shrubs and plants we observe the voracious spider, watching for its unwary prey on or near its beautiful snare; and flying around are numerous winged insects of the most varied forms and habits, some searching for suitable spots on which to deposit their eggs, some visiting flowers for nectar, and others seeking out decomposing matter from which they extract their food, or in which they recognize a store of food for their offspring.

The above are but a few of the many interesting objects on which observant children may direct their attention, either in the school garden or further afield. In fact, the lower forms of life—life belonging to the invertebrate world—are so numerous and varied and withal so interesting, that they alone will give ample employment for the most enthusiastic observers throughout the whole summer. And since the creatures we have mentioned, together with many others, are to be found in every garden, even in the very heart of a populous town, no teacher need despair of a lack of material for valuable nature lessons, even though nearly all the higher forms of animal life are almost entirely wanting.

Ponds and streams are also teeming with life during the summer months, and a few dips with a gauze net into a weedy pool will supply material for the school aquaria sufficient to keep all the little observers busy till the end of the season.

The school aquaria should now be well-stocked, with due regard to the separation of the carnivorous from the harmless species, as laid out in our chapter on the aquarium and its management.

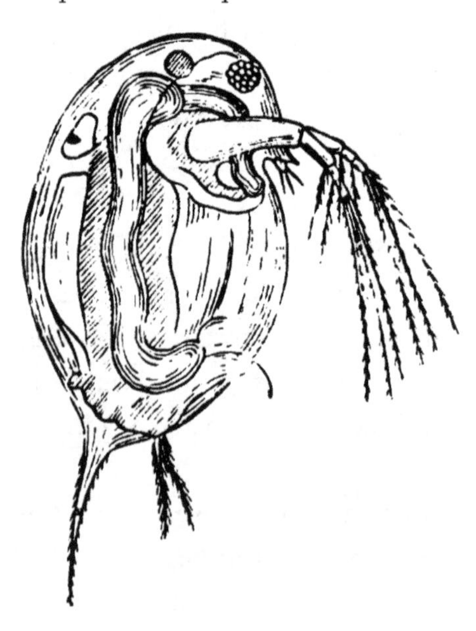

A Water Flea (Fig. 132)

When pond material is being collected, it is probable that the majority of teachers will find among the miscellaneous forms acquired several species that are quite unknown to them. Such forms need not be rejected simply because they are unfamiliar. One may become thoroughly interested in a natural object, and learn much about it without even knowing its name. We have previously said that all school nature study should be carried on in the spirit of research; and why not let this be the spirit of the teacher as well as of the class?

A child may ask the names of the creatures under observation, but no teacher need be ashamed because he is unable to name correctly all the objects that may be found from time to time. Let the teacher and the children work together, each endeavoring to discover as much as possible of the movements and other habits of the creatures observed, and to solve the many interesting and instructive problems connected with the structure, growth, metamorphosis, habits, etc., of these creatures.

The study of pond life must always prove interesting, if only on account of the great variety of forms among aquatic creatures and their equally diversified habits. The following are a few of the common forms of pond life eminently suitable for the observation of children, together with a few hints as to the chief features to be observed.

1. Water fleas, water mites, cyclops, hydras and other minute creatures which, although distinctly visible to the naked eye, are more advantageously observed with the aid of a magnifying lens.

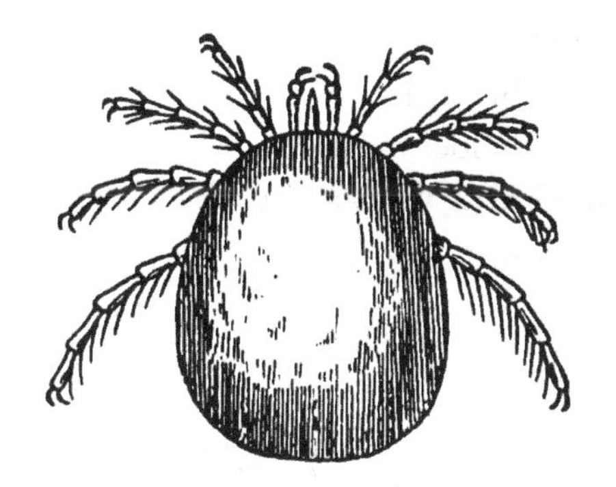

A Water Mite (Fig. 133)

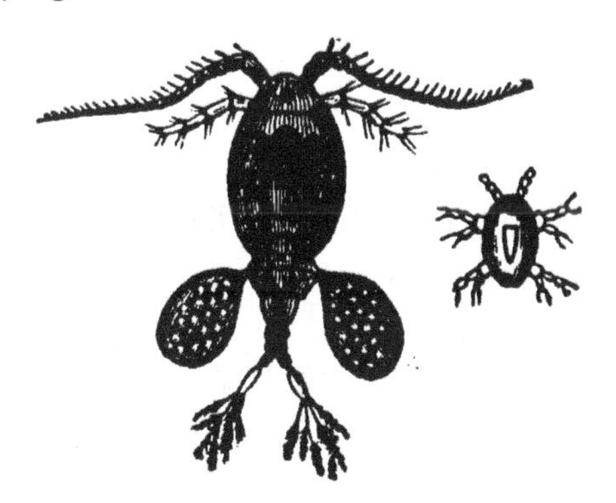

A female Cyclops with Egg sac (Fig. 134)

These are, of course, more suited for the observation of the senior children. They are more conveniently examined in small vessels, such as tumblers, small glass flasks, etc., to which they may be transferred for the purpose—the water fleas, cyclops, etc., by means of a pipette, and the hydra with the pondweeds to which they are attached. Let the children note how the water fleas and cyclops (which are not insects, but minute crustaceans) differ from insects in structure (assuming that they have previously been made acquainted with the leading characteristics of insects), and let those who have previously studied the life of the rock-pool on the coast compare the hydra with its near relative—the sea anemone.

2. Leeches. Note their peculiar method of locomotion, and compare their ringed bodies with those of earthworms.
3. Water snails. Of these there are several common species with shells of different forms. Their movements are readily observed as they creep on the glass sides of the aquarium, or, in an inverted position, glide along beneath the surface film of the water. They may also be seen feeding on the green encrustations that form on the glass of a well-lighted aquarium, the movements of the mouth being distinctly visible when viewed through a lens; and the development of the young snails from the eggs, which are embedded in a jelly-like covering, may also be minutely observed.

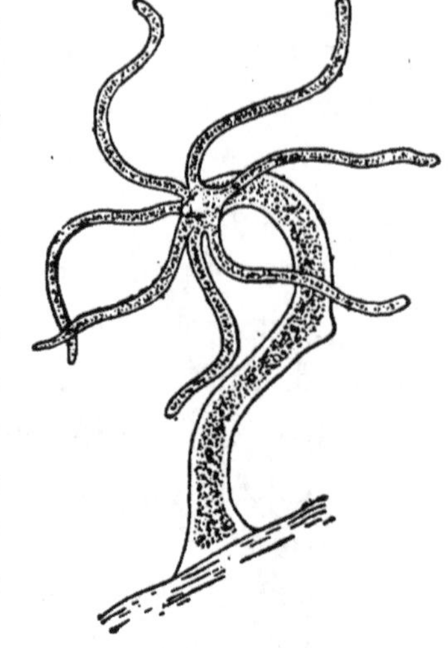

The Hydra (Fig. 135)

4. Water spiders. These are very interesting creatures. Among other things, the children will observe the construction of sub-aquatic threads; the making of the diving bell, and the mooring by which it is secured to the pondweeds; also the means by which the spider carries down its supply of air, to enable it to live beneath the surface of the water where it watches for its prey. (See page 69.)

The Common Horse Leech (Fig.136)

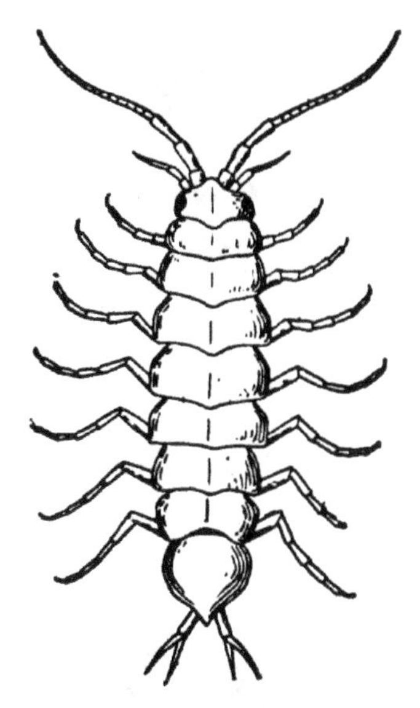
The Water Hog-Louse (Fig. 137)

5. The water hog-louse—a creature that somewhat resembles the common woodlouse which is so abundant in our gardens. It is not an insect, but a crustacean. Its movements are interesting, and it is very useful in the aquarium, where it devours decomposing matter, and is, therefore, valuable as a scavenger.
6. Larvae of caddis flies. These, too, are very instructive. There are several species, varying in size, but all similar in structure and habit. Their bodies are not protected by a hardened skin, with the exception of the head and the next three segments, and they protect themselves by constructing a cylindrical case which they drag about as they move from place to place. The different species employ different materials for this purpose. When fully grown they change to four-winged flies.
7. Water-beetles and their larvae. These are mostly carnivorous insects, preying on the harmless or less predatory creatures of ponds and streams, but their movements are particularly interesting, and they afford striking examples of adaptation of structure to habit.
8. Larvae of dragon flies—also very predatory, and provide special opportunities of witnessing interesting examples of insect metamorphosis.
9. Various other aquatic insects are to be found commonly in ponds and ditches—all more or less interesting and instructive, some undergoing but little change, and others exhibiting remarkable changes of form.

The children of schools situated near the sea should continue to study the life of the rock-pools left by the receding tide; and, if the school is of such a distance from the coast that but few opportunities of rambles on the beach are possible, some of the creatures observed should be taken alive for further examination in the sea-water aquarium. Such studies may include, at least, the following: Sea anemones, starfishes, sea-urchins, various marine mollusks and their shells, marine worms and their homes, crabs, shrimps, prawns, the fishes of the rock-pools, and the principal birds which inhabit the shore.

Caddis Fly Larva, their cases, and an adult Caddis Fly (Fig. 138, 139, & 140)

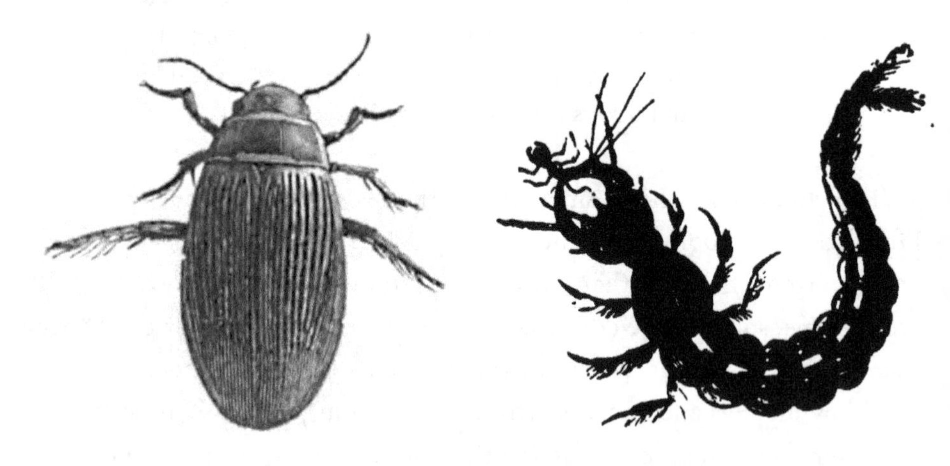

The Great Carnivorous Water Beetle and its larva (Fig. 141 & 142)

Whatever be the animals selected for study, special attention must be called to their movements; and the children should be encouraged to endeavor to distinguish between those movements which are apparently instinctive and those which show signs of intelligence. Whenever necessary, the teacher will stimulate and guide the observations, and also encourage the solution of problems connected with the phenomena observed, by asking questions of the children.

Thus: How does the creature defend itself from its enemies? or, How does it attack its foes? How does it capture its prey? or, How does it obtain its food? How is the food eaten? Does it construct any kind of shelter or home? and, if so, What is the nature of its home, and how is it constructed? Does the creature live a solitary life, or is it gregarious or social in its habits? If the latter, What is the advantage of the social life? How do the creatures employ themselves in their homes? Is there any division of labor among them? Does the creature store food not required for immediate use? If so. What is its food? How is it obtained? and How is it stored? Does the animal know and care for its young? Does it prepare a home for its young, protect them, feed them, or teach them how to obtain their food? Does it construct a snare for the capture of its prey? If such is the case, What is the nature of the snare, and how is it constructed?

Dragonfly larva & adult (Fig. 143)

Particular attention should also be called to instances of resemblance to environment and of mimicry. It is often erroneously supposed that such instances are to be found almost exclusively among tropical and other exotic species, but we need go no farther than our own garden and the neighboring ponds to find numerous instructive and highly interesting examples.

Some very common caterpillars that feed on our garden plants and on our forest trees strikingly resemble the twigs to which they attach themselves when at rest. Some of these are green, and repose on green twigs; while others which rest on brown twigs are colored accordingly. The resemblance is rendered still more remarkable by the position assumed by the larva when at rest, for it fixes itself firmly by means of its hindmost appendages (prolegs and claspers) with its straightened body standing out at an angle from the twig; and, not infrequently, we find the body of the caterpillar bearing color-patches and projections that closely resemble the leaf-scars and other markings of the twig on which it is situated. Some green caterpillars, too, including one that is commonly found on cabbages in kitchen gardens, apply themselves so closely to the underside of leaves that they look just like projecting veins.

A large dragon-fly (Fig. 144)

It is obvious that the resemblances above referred to are of a protective nature, for the caterpillars are much sought after by birds and other insect-eaters; but equally interesting are those cases in which predatory creatures, with a close resemblance to their environment, are enabled to lie concealed when watching for their prey.

Note, for example, the sluggish water scorpion—one of the carnivorous water bugs, the back of which is mud-colored. This insect is often to be seen in very shallow water, resting with its flat body on the mud so that it is not noticeable except when very closely examined, and ready to seize any harmless creature that comes within reach of its forceps-like front legs. Another voracious water bug, known as Ranatra, also very common in ponds and ditches, is not only of the color of the mud on which it rests, but, with its legs applied closely to its elongated body, looks just like a piece of dead stick.

The above are a few of many examples of resemblance to environment to be seen among common British species. Call the attention of children to one or two such instances, and they will soon discover others. Moreover, the various examples met with will display such a diversity of features that each new discovery will probably present some fresh feature, and open up some new line of thought as regards the problem to be solved.

The Water Scorpion (Fig. 145)

1. With fore legs 'closed.'   2. With wings extended.   3. Larva

Ranatra (Fig. 146)

Concerning the study of vertebrate animals, little need be said beyond the remarks made in connection with the spring season. The movements of small fishes may still be observed at times in the school aquaria, where new species should be introduced as they can be found. Frogs, toads, and newts (including the young ones) have now left the water to seek their food and shelter on land. A few of these may be kept as pets in the vivarium, providing they are given sufficient room, and kept under conditions approaching, as nearly as possible, those which obtain in Nature, with due regard to proper feeding; but these creatures are better observed in a state of semi-captivity in the school garden, where they will find their own food, a shelter being provided for them by piling up some stones in a shady corner.

The common British reptiles may be seen sunning themselves on banks or in the open, and there is no reason why they should not be kept as pets, providing they are properly cared for.

The birds inhabiting the neighborhood of the school may still be encouraged by scattering their favorite foods in the playground, or by placing daily supplies on feeding tables in convenient situations for observation. Mixed birdseed, and ripe thistleheads, teasel-heads and hard-heads (knapweed) will prove very attractive to the hard-billed finches and other songsters, while bread-crumbs, and bread moistened with milk will entice many of the soft-billed species. A small vessel of drinking water, refilled every day, will also attract considerable numbers, especially during periods of drought; and a shallow vessel of water placed on the ground will enable us to watch the interesting movements of the birds as they enjoy their bath.

Many of the habits of birds may be studied from tame, caged creatures kept in the school, but the birds so kept should always be those which have been reared in captivity. Wild birds that have been trapped should never be imprisoned.

As regards wild mammals, the children living well within towns and cities will have but few, if any, opportunities of observing them, but those children who live on the outskirts of towns may find much interest in watching the movements of such little creatures as the field vole and the bats. Country children will see several other species, especially during harvest-time, when so many living things are roused from their hiding places among the corn and other crops. Whenever possible, the little mammals discovered should be traced to their homes, the nature of their hiding places observed, and their various habits noted.

D. Physical Studies

The studies of the earth, air and sky for the summer months will not differ materially from those of the spring. The children may continue to observe the positions of the rising and the setting sun on various days as long as these events take place at hours during which they should be awake, and to keep records in their notebooks of the times of rising and setting, together with the lengths of the days.

They will note the gradually increasing temperatures as the season advances, observing that our hottest days are not those which are the longest, but that the accumulation of heat continues after the longest day is past—that as long as the days considerably exceed the nights in length, there is an accumulation of heat due to the fact that more heat is gained during the day than is lost, by radiation, in the night.

The senior scholars should still continue to observe the altitude of the sun at midday, with the aid of the simple apparatus previously described (page 97), special attention being given to such observations towards the end of June, when the altitude of the sun is at its maximum and remains practically the same for several days together.

The elder children will also be interested in the movements of the shadow, cast by the sun, of an upright, thin rod fixed in the center of a large compass card placed in its proper position as regards the geographical points. It will enable them to discover that the sun is not always exactly in the south at noon, and that, as a consequence, the time as read from the sundial does not always agree with that indicated by a correct clock. Simple sundials may also be constructed by the children, and observations made on them from time to time. These dials need only consist of upright rods (styles) mounted on thick, white cards, or on slabs of whitened wood; but where the children are sufficiently skilled there is no reason why they should not be taught to construct a horizontal dial, with style parallel to the earth's axis—making an angle with the plane of the dial equal to the latitude of the place; or a vertical dial, in which the angle formed by the style and the dial-plate is the complement of the latitude.

The study of the face of the sky during the summer months is not likely to occupy much time, since the very long days preclude the observation of the heavenly bodies, at least to a considerable extent. The senior scholars, however, may be encouraged to watch the movements of planets and stars, giving special attention to any of the former that may have recently appeared, and to a few of the more conspicuous constellations of stars that were not visible earlier in the year.

Finally, attention should be called to the climatic conditions of the season. The readings of thermometers may be taken regularly, and the records preserved for comparison with those of other seasons as well as with those of subsequent summers. Such readings may include those of the maximum and minimum thermometers, sun and shade temperatures, and indoor temperatures. Summer showers and droughts should be studied, more particularly in connection with their effects on vegetable life; and the children of rural districts will be encouraged to note how the weather conditions affect the quantity and quality of cultivated crops.

— 6 —

Autumn Studies

A. General Remarks

We have now reached the season characterized by a more or less rapid decline of life. The sun is becoming lower and lower in the sky day by day, and, more heat being lost during the gradually lengthening nights than is gained during the shortening days, the loss of this energy becomes increasingly apparent.

Flowers are rapidly becoming less abundant, and the withering and falling leaves of herbaceous plants reveal the bare stems that bore them. Flowers give place to ripened fruits and seeds which are destined to assist in the renewal of life in the following spring. The trees and shrubs are making preparations for their winter rest, but not until they have formed the buds that are to develop into the branches of the following year, and scattered the seeds that are to produce their offspring.

Animal life is also on the decline. Many creatures have already entered into their winter condition, even early in the season, and others are making preparations for approaching winter. The number of living beings is rapidly reduced, later in the autumn, by the equinoctial storms and early frosts, and all appearances point to the coming of a period of a minimum of energy and life.

B. Vegetable Life

I. The Ripening and Dispersal of Fruits and Seeds

Although, during the autumn, the wildflowers are rapidly disappearing, this season provides an abundance of material for very interesting studies.

Many of our plants and trees have produced and ripened their fruits during the summer, and some even in the spring, but autumn is undoubtedly the best season for a special study of the different kinds of fruits and of the means by which both fruits and seeds are dispersed.

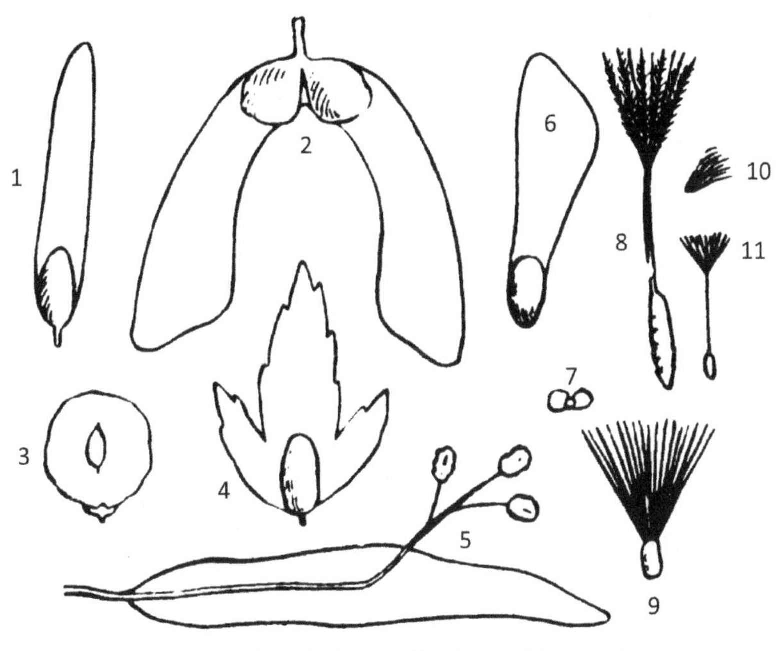

Fruits and seeds dispersed by the wind (Fig. 147)

1. Ash
2. Sycamore
3. Elm
4. Hornbeam
5. Lime
6. Pine
7. Birch
8. Goatsbeard
9. Thistle
10. Willow Herb

All except the youngest children should have a clear idea of the true nature of a fruit, and be able to distinguish between fruits and seeds, even when the former are distinctly seedlike in their appearance. They should be taught to observe certain flowers, from time to time, in order that they may watch the formation of the fruits—the gradual ripening of the ovaries. Such observations will afford suitable occupations for school rambles in the open country, where this is possible; but town children are placed at no great disadvantage in this respect, for the development of fruits may be studied most satisfactorily in quite a large number of easily grown, cultivated plants, and equally so in the case of many of our commonest weeds.

During autumn rambles there are splendid opportunities of collecting fruits and seeds of various kinds for detailed study at home or in the schoolroom, but the modes of dispersal are almost essentially outdoor studies. On a breezy day in autumn, wind-dispersed fruits and seeds are to be seen sailing through the air at almost every step; and while the children are at work among the hedgerows, along weedy waysides, in woods, and in the open meadow, a variety of hooked, bristled, and barbed fruits will firmly attach themselves to their clothing, thus providing them with material for studying dispersal by the agency of animals.

Birds may also be seen at work among ripe succulent fruits, devouring the edible portions, and rejecting the 'stones' which are often removed to some distance from the tree or shrub on which they grew; or, in the case of berries and similar small fruits, swallowing them whole, and afterwards depositing the indigestible seeds they contained with their excrement.

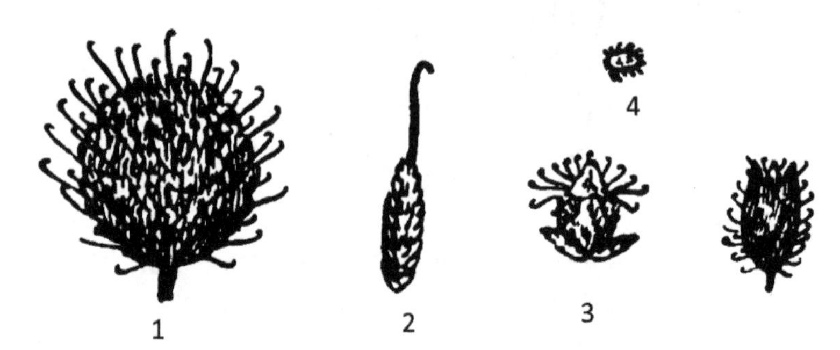

Fruits dispersed by the agency of animals (Fig. 148)

1. Burdock
2. Avens
3. Agrimony
4. Wild Carrot
5. Forget-me-not

Again, on a bright, sunny day, the crackling sound produced by the sudden, elastic splitting of fruits may be heard almost continuously in certain spots; and a little patient searching and watching will enable the scholars to witness the mechanical distribution of the seeds these fruits contained. Some of these fruits, such as those of the broom, vetch, wild geraniums, balsam, etc., may be taken home in a ripe condition, but as yet entire, and placed on a large sheet of white paper or a spread table-cloth in a sunny situation, in order that the mechanical dispersion of the seeds may be more closely observed, and the distances to which they are thrown may be measured.

Another method of mechanical dispersion may be observed in the fruit of the common dog violet. The fruit of this plant splits into three parts which separate widely but still retain their seeds. As these parts become dry, however, their sides become straightened, thus pressing against the seeds, and shooting them out one by one just as one might shoot an orange pip by pressing it between finger and thumb. This may be observed at home by placing a few of the fruits on a sheet of white paper exposed to the sun or spread before a fire.

The Great Willow Herb (Fig. 149)

The above are only a few examples of the many and varied methods of dispersal; and during the present season a large amount of time may be profitably spent in studies relating to this subject, attention being given not only to herbaceous plants, but also to the common forest trees and hedgerow shrubs.

In those town schools so unfavorably situated that country rambles can seldom or never be attempted, much may be done with the aid of plants grown in pots, boxes, or in the school garden. The plants named above, together with the dandelion, thistle, Michaelmas daisy, great willow-herb, and various other species, that display interesting provisions for the dispersal of their seeds, will provide very useful material for study.

Hitherto we have spoken of fruits and seeds with special reference to their modes of dispersal; but, apart from this consideration, much interesting work may be done in the autumn by observing the different forms of dehiscent fruits and the manner in which they open to set free their seeds. Encourage the children to collect various kinds, either in the open country or in their own gardens, and then, with the teacher's aid when necessary, to classify them according to the manner in which they split.

This portion of the study of fruits will certainly help to enlarge the children's experiences with regard to the subject of dispersal, for, in the collection and examination of splitting fruits, they will frequently meet with specimens that open suddenly, and with such force that they throw their seeds a considerable distance, many being so sensitive that they readily burst as they are handled. Thus, the little triangular fruits of the shepherd's purse, when quite ripe, will burst asunder with the slightest touch.

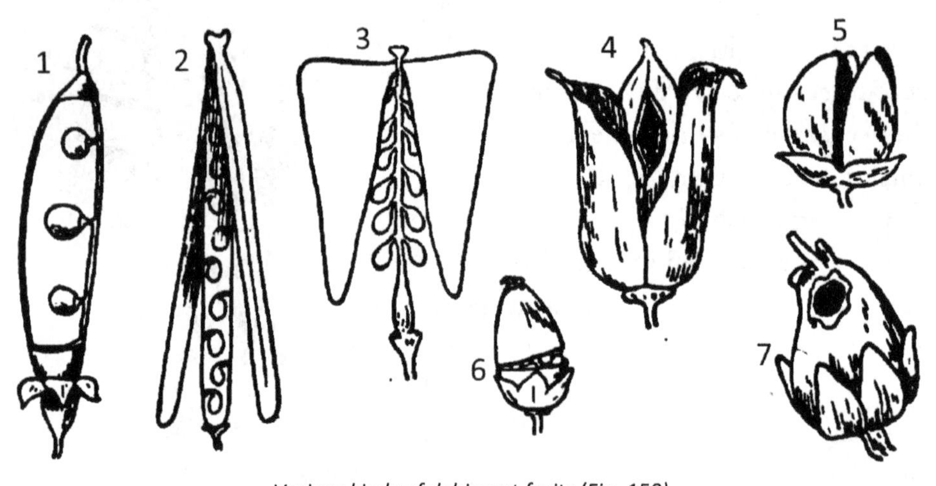

Various kinds of dehiscent fruits (Fig. 150)

1. The Pea
2. Wallflower
3. Shepherd's Purse
4. Larkspur
5. Mullein
6. Plantain
7. Antirrhinum

The following list of very common dehiscent fruits may serve as a guide to the teacher, more especially as a means of assisting him in the selection of suitable specimens for class work in the schoolroom:

1. Legumes or pods—fruits that split into two parts, having the seeds attached down one side only—pea, bean, vetches, bird's-foot trefoil, broom, furze, etc.
2. Siliquas—fruits in which two valves separate from a central membrane to which the seeds are attached—wallflower, rocket, cresses, shepherd's purse, pepperwort, candytuft, mustard, stock, cabbage, etc. (The short and

broad fruits of this description are generally distinguished by the name of siliculas.)

3. Follicles—fruits which open on one side only—larkspur, columbine, peony, etc.
4. Capsules—fruits which split, when ripe, in various other ways—
    - a) Splitting longitudinally—mullein, St. John's-wort, lilac, figwort.
    - b) Splitting transversely—scarlet pimpernel, plantain.
    - e) Splitting by the separation of teeth—campion.
    - d) Opening by means of the formation of pores or slits—snapdragon, campanula, poppy.
5. Fruits splitting into two or more one-seeded parts—sycamore, mallow, hollyhock.

After the children have examined and classified the fruits that have been collected, including the various succulent kinds, they may be encouraged to inquire into the uses of the different cases or coverings which enclose the seeds. Some of these uses have already been considered, inasmuch as they are connected with modes of dispersal; but, in addition, it should be made clear that the harder and more indestructible fruit cases, since they decay very slowly, serve to protect the enclosed seed or seeds, and to prevent germination, until the most favorable season for development has arrived; also that the fleshy, succulent coverings often produce, as they decay, a small amount of plant food which helps to sustain the seedlings during their earlier stages of growth.

Furthermore, the autumn collection of fruits and seeds will supply much suitable material for the study of the development of young plants in the following spring. A selection of them should be preserved for the purpose mentioned. When the spring arrives, allow the children themselves to select, as far as it appears advisable, the seeds they would like to grow, for they would naturally take more interest in the germination of those species which had previously aroused the greatest curiosity, either on account of some peculiarity of the fruits which contained them, or of the plant which produced them.

II. Autumn Flowers

Little need be said concerning the autumn flowers, since hints have previously been given concerning the methods to be adopted with regard to the study of flowers in general. The more conspicuous autumn blossoms should be noted, particular attention being given to the habitats and habits of the plants to which they belong. It will be observed that a very large proportion of these autumn flowers are composites—members of the daisy order.

The localities and habitats producing the largest number of flowers should also be noted, and an endeavor should be made to ascertain why these situations are more prolific. Some of the summer blossoms will be found to linger well into the autumn season in certain spots, and second crops of summer flowers, often much after their normal season, will frequently be found on banks and hedgerows that have been cleared with the sickle, new and late growths having been produced on the old stumps of plants that have been cut down.

III. Autumn tints & leaf decay

This is undoubtedly one of the most interesting studies for the present season. The children should be taken to a neighboring wood, if possible, or to a wooded park or open space in or near the town, where they can study the beautiful autumn colors of the trees and shrubs; and even the hedge bank, covered only with herbaceous plants, will afford a certain amount of variation in the color of the foliage due to the autumnal changes.

It will be found that some of the children show but little interest in the wonderful display of color before them, but the teacher who himself derives pleasure from such a scene will soon be able to arouse some enthusiasm among his apathetic scholars by calling attention to those general effects and more detailed features which appeal to him. His own enthusiasm, expressed not only by his manner, but also by a language inspired by an ardent appreciation of the scene, is sure to have a good effect on the scholars, awakening the interest of those who have already learned to appreciate the beautiful, and helping to cultivate an aesthetic taste in those who, as yet, show no signs of such an appreciation.

Let the children gather some twigs displaying the most beautiful of the autumn tints, take them home or to the school, and attempt to represent them with the aid of brush and colors. Let them also attempt to describe them in words, and write a description of the scene witnessed as they stood looking on the wood. Thus they make use of Nature as a means of training the eye, the hand, and the mind, and also as a means of giving expression to their thoughts in words.

But then, the study of autumn tints also affords much material for thought, especially when considered, as it must be, in connection with the falling of the leaves. To a thoughtful child the following questions, or, at least, some of them, may present themselves:

1. Why should the leaves change color before they fall?
2. Is there any connection between the change of color and the fall of the leaf?
3. How is it that the leaves of different trees and plants change to different colors? Why should the willow leaves turn grey, those of the birch a pale yellow, those of the oak brownish yellow, the aspen orange, the alder brownish green, the dogwood violet, the service tree purple, and the mountain ash scarlet?
4. Why should the leaves fall at all? Is it necessary for the welfare of the trees themselves?
5. What causes them to fall? etc.

Of course the answers to some of these questions are too advanced for the younger children, but all may be explained more or less completely to the senior scholars, especially those who have an elementary acquaintance with the nature of chemical changes.

A full account of the preparations made by deciduous forest trees (and the same applies to many herbaceous plants) for their winter rest would occupy much more space than we can give, but the following notes embody some of the principal facts connected therewith. As far as possible, the facts should be elicited from the children by a proper presentation of questions leading them into the correct line of thought, but the teacher may often find it necessary to come to their aid with statements of facts which they are unable to give, but which lead to further problems that the children may attempt.

1. Trees, other than evergreens, must necessarily cast their leaves on the approach of winter, because these leaves are too delicate to withstand the winter storms and frosts. Even if they could survive the winter, and remain functional, they would be a source of so much loss of moisture at a period when the roots are inactive (on account of the low temperature) that they would endanger the life of the trees.
2. The leaves contain much valuable material which the tree cannot afford to lose. Hence this material, in a changed condition, is absorbed into the branches and trunk, to be stored for the following season, before final arrangements are made for the shedding of the leaves.
3. Among other materials, the green coloring matter (chlorophyll) of the leaves is dissolved and absorbed to a greater or less extent.
4. The chemical changes connected with the transformation of the compounds which have to be absorbed give rise to byproducts, including a blue substance called anthocyanin, and certain acids.
5. Anthocyanin is turned red by acids; and the colors of autumn leaves are produced by the extraction of chlorophyll, and the formation of anthocyanin and the action of the acids thereon; the different colors and tints depending upon the extent to which these products (and possibly several others) are formed and to the proportion in which they are present in each case.
6. While these changes are taking place, a separation layer of cork cells is being formed at the base of the leafstalk or in some other situation, the cells gradually extending from the circumference inwards until the layer is complete. The walls of the cells soon become so brittle that the leaf is easily detached by the wind.
7. Leaves fall very rapidly on the approach of the frosts of late autumn, but, though it will be understood from the preceding note that frost is not the cause of leaf-fall, yet the frost, by increasing the brittleness of the separation cells, considerably accelerates the rupture.
8. Finally, although the leaves are reduced practically to dead skeletons by the extraction of most of the nutritious matter they contained, yet they contain, when they fall, a certain amount of material which, by its decomposition, adds to the fertility of the soil. The leaves of evergreens are of a firmer texture than those of deciduous trees, so that they are not

so easily destroyed by frost. They have also a thicker epidermis, and thus retain their moisture better; consequently they do not allow that loss of moisture which would be detrimental to their plants or trees at a time when the roots are inactive on account of the low temperature of the soil. Yet the leaves of evergreens are shed, but the fall is not conspicuous since it takes place more or less throughout the year, the lower leaves becoming detached as they die, and their function being taken up by the new leaves produced at the tips of the branches.

Children should be encouraged to note the dates on which the fall of the leaf commences with different trees, and the times when the fall is complete and the trees are bare. Such dates entered during one year should be compared with those of another, together with notes on the weather at the time or immediately preceding. Thus the children will become acquainted with the conditions which hasten or which retard the leaf-fail.

They will notice, too, that the time of denudation varies with the same species of tree in different situations, and endeavor to determine what conditions of soil and situation are most favorable to the retention of the leaves.

Furthermore, they will be interested in observing that while in some species of trees the bases of the branches become bare first, so that the leaves at the tips of the twigs are the last to fall, in other species the reverse is the case.

As regards biennial and perennial plants, both teacher and scholars should be on the look-out for instances of the storage of food for the growth of the coming year. Such stores are often to be met with in the thick rootstocks which survive the winter, in thick underground stems, tubers, bulbs, etc. Some of these storage parts may be taken home for cultivation in the garden; or the earlier stages may be studied in the following spring by growing them in water or in wet sand.

During the autumn studies of the vegetable world we see everywhere the natural decay of those structures, now dead, that have completed the performance of their functions. Leaves are falling and rotting on the ground. The old stems, now no longer required, are gradually wasting away, the moisture evaporating, and the softer tissues breaking up, till practically nothing remains save the less destructible fibers.

The senior children may be taught the true meaning of decay. Let them have a simple account of the minute bacteria, some of which are so small that a mass of several millions would weigh only about a grain; and give them to understand that the growth and multiplication of these minute cells is one of the principal causes of the decay of vegetation, and that they are the sole agents concerned in the decomposition of animal substances. They may further learn that, as a result of the action of these organisms, all dead matter is broken up, some of its substance passing off into the air as gases, while the remainder, changed in its composition, becomes incorporated into the soil; and that the surface of the earth is thus kept fresh and unencumbered for renewed growths.

In addition to all the autumn studies mentioned above, all children, and more especially those of village schools, should be encouraged to interest themselves in the various human activities of the season: the gathering in of crops; the ploughing of the soil, manuring, and other preparations for future crops; the sowing of seed for the produce of the coming year; the storage of fodder of various kinds for the use of cattle during the winter, etc.

C. Animal Life

There is also a gradual decline in animal life as the autumn advances, and this is most noticeable in the case of insects and the various small invertebrate creatures that form such a conspicuous feature of the country and the garden during the warmer months.

None of the species, however, disappear entirely without having previously made some provision for their perpetuation. Butterflies, moths, and various other insects deposit their eggs before they die, and these eggs are to be found on the bark or the twigs of trees, on fences, and in various sheltered places.

Other insects, dying earlier in the season, produce their eggs so much earlier that the larvae have hatched out and made some progress in their growth before they settle down for a period of hibernation. Such larvae may be seen concealed under the cover of thick herbage, among fallen leaves, or beneath the surface of the soil. They climb their food plants and feed as long as the weather conditions remain favorable, but, as the colder days of late autumn set in, they return to their hiding places to remain till the following spring.

Quite a large number of insects have now reached the more advanced stage of the pupa or chrysalis, and numbers of these may be turned out of their quarters in the crevices of the bark of trees, under the soil, or in some other cover. Some are enclosed in silken cocoons, some suspended by silky threads, others in cocoons or cells molded in the earth or constructed by gluing or binding together various materials.

Many of these chrysalides will be found as we turn over the soil of the garden during the autumn, each one being generally close to the roots of the plant on which it fed as a caterpillar. Large numbers, too, may be dug out of the angles of the roots of trees, the caterpillars, when fully grown, having descended the trees on which they lived, and then buried themselves for the winter in the first suitable spot.

The chrysalides may, of course, be found also throughout the winter and early spring, but autumn is the best time for collecting them, since large numbers are devoured by moles, beetles, and various other insectivorous creatures. If they are collected, and preserved until the spring, they afford very favorable opportunities for studying the perfect insects, including many species that are seldom seen on the wing by ordinary observers, for the majority of the chrysalides so obtained are those of moths that fly at dusk or during the dark hours of the night. They will also provide opportunities for the observation of the emergence of the winged insects, and of the gradual expansion and stiffening of the wings.

It must be remembered, however, that if pupae are to be preserved alive until the time of the appearance of the perfect insects, they must be kept under conditions corresponding as nearly as possible with those under which they were found. Those that were obtained above ground or in very dry places need only to be placed in shallow boxes, on a layer of moss, sawdust or bran, and set aside for the winter in a cool room, out of the reach of mice; while those dug out of the soil should be covered with a little sifted soil, with a layer of moss above, and occasionally damped with a slight sprinkling of water.

Silken cocoons, spun by spiders, containing a large number of little white or yellow eggs, may be seen in abundance during the autumn, on fences, in the crevices of the bark of trees, and in other sheltered places. If a few of these be collected, and preserved in small, wide-mouthed bottles with a piece of muslin tied over the top, they will yield numbers of young spiders in the following spring and give the children an opportunity of learning that spiders are perfectly formed from the first—that they do not undergo metamorphosis similar to those which are characteristic of the majority of insects. After the young spiders have been examined, together with the silken web which they construct, they may be set free in the school garden, where, in the summer, when they are fully grown, they will construct their snares and afford splendid opportunities for the observation of their habits.

Other small creatures commonly found in gardens, hedgerows, etc., such as snails, slugs, earthworms, centipedes, woodlice and ground beetles, may be studied in the autumn as earlier in the year; but special attention should be given to the preparation made by these, towards the end of the season, for their winter rest. They may be traced to their hiding places, and the nature of their winter homes ascertained. Much interest will be found, for example, in the discovery of a colony of snails that have congregated in a suitable cozy corner, in the manner in which they have closed their shells with a membrane formed of hardened slime and securely cemented their shells to the surface on which they are to be fixed for their long, winter sleep.

The observation of the movements of certain birds will also give interesting occupations for the season. The summer visitors which have been with us for several months now collect together previous to their departure for southern climes, and the winter visitors that have been breeding farther north begin to arrive. In addition to these there are the birds of passage which have their breeding-grounds outside the limits of our islands, but make a brief sojourn here on their way, probably for rest and food.

D. Physical Studies

The studies of the earth, air, and sky for the autumn will not differ materially from those of the other seasons. Of course attention will be drawn to the rapidly shortening days, and the variation in the positions of the rising and the setting sun day by day; also to the decreasing altitude of the midday sun and the corresponding decrease in temperature.

The nature of autumn gales will be noticed, and their effects in hastening the defoliation of the trees; also the denuding action of equinoctial storms, the swelling of streams and rivulets, and the formation of wet-weather streams in small hollows that may have been dry for some period.

Simple explanations of the formation of mists and fogs may be given, and, where possible, these explanations should be illustrated by means of experimental demonstrations of the processes of evaporation and condensation, and of the various forms assumed by water under changing conditions.

As before advised, attention may also be directed to the planets visible at the time, if any, and to a few conspicuous constellations of stars that were not to be seen earlier in the year.

— 7 —

Winter Studies

A. General Remarks

During the winter life is at its lowest ebb. Many animals are in a dormant condition, spending the whole of the season in a state of perfect repose, and taking no food. In fact, in many instances their natural food is not now to be found. Others are winter hiders, retiring to their sheltered nooks, and remaining there as long as severe weather persists, but emerging into the open in search of food during milder intervals.

As regards vegetable life, although most plants and trees are in a resting condition, exhibiting no signs of growth, yet some, and especially those which flower very early in the spring, or even before the winter is over, are making some progress and provide interesting studies for the season.

The winter condition of our common trees and shrubs will, of course, now receive special attention, and these will give abundant opportunities for both indoor and outdoor work.

During this season, too, time will probably be most easily found for the observation of lifeless natural objects, such as the common rocks, building-stones and paving-stones of the neighborhood; also favorable opportunities will present themselves for the study of the special phenomena of the cold season, such as frost, ice, winter storms, snow, etc.

B. Plant Life

I. Winter condition of trees and shrubs

The study of the winter aspect of deciduous trees and shrubs is a most valuable one, affording plenty of work both outdoors and in the schoolroom; and while the children living in the country or on the outskirts of towns are well-favored as regards this branch of nature study, more especially as far as the outdoor work is concerned, yet town children will find plenty to do, since a very large proportion of our common forest trees and hedgerow shrubs are to be found in the parks and open spaces of nearly every large town. It is always better to make a study of a tree as a whole before particular attention is given to its parts.

An occasional ramble during fine weather is most important, since it gives splendid opportunities for noting the general appearance of the winter condition, and for comparing this with the memory pictures, or sketches, made in the summer while the same trees were in full leaf. Let the children take particular notice of the general form of each tree from a distance and, if the weather is not too cold, make a rapid sketch of its appearance.

The Scots Pine (Fig. 151)

Then, at a nearer approach, but still at a little distance, let them note the mode of branching—the general disposition of the branches and the peculiarities of the twigs. Next, approaching close to the tree, they study the size of the trunk, the nature of the bark, and the form and arrangement of the buds. Finally, a twig is removed for a closer study on the return to school or home.

The Yew in Flower (Fig. 152)

Particular attention should be paid to the arrangement of the buds, and to the relation between this arrangement and the mode of branching. We have already dealt with this subject as one suitable for study during early spring, while the trees are still bare, and there is no need to repeat here the hints previously given.

In fact, many of the studies mentioned in Chapter 4 as suitable occupations for the early spring, are really those appertaining to the whole of the cold period of the year; and, during the bleaker winter months, any of these studies for which time could not be found in the spring may be taken up.

Evergreens may also be studied now with advantage, and their principal characteristics noted, unless this work has already been done during the spring or autumn.

The British conifers—the Scots pine, yew and juniper—together with some of the more abundant exotic species of this group of trees and shrubs, will be included under the heading of evergreens, excepting the larch, which is bare in the winter. The flowers of these trees and shrubs generally appear after the winter is over, but their general forms, branching, leaves, and cones, will provide interesting studies for the present season.

Some of the pine cones will be found with their winged seeds still between the woody scales, and, after examination, a few of these may be preserved for the study of their germination when the weather becomes a little warmer.

If an opportunity for a ramble through a pine-wood or larch-plantation occurs, a special study of the general features of, the wood will be of interest. The dense covering of pine-needles will be observed, the old cones from which the winged seeds have escaped, and frequently the little heaps of broken cones in which nearly all the woody scales have been separated from the central axis by the busy squirrel in order to obtain the seeds for food. It will also be noted that the ground is often entirely devoid of green vegetation beneath the pine-trees, and the children will endeavor to discover the cause of this peculiarity.

II. Hedgerows, waysides, etc., in winter

The Shepherd's Pulse (Fig. 153)

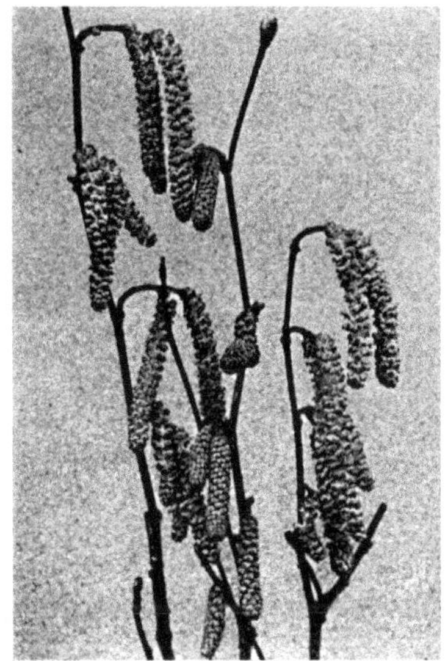

The Hazel in February (Fig. 155)

While a large number of hedgerow and wayside plants lose all their leaves in late autumn, many retain a certain amount of green foliage throughout the winter, especially in sheltered situations. Field and moorland are also covered with green, and although the color is due principally to various species of grasses, yet quite a variety of interesting, hardy plants will be found among the blades, most of them, however, being of such a low growth that they are sheltered effectually by the latter.

Even flowers are to be seen throughout the winter, including those (shepherd's-purse, duckweed, groundsel and red dead-nettle) which have already been referred to as blooming throughout the whole of the year, at least in sheltered spots. Then, before the winter is over, the wild snowdrop will be seen in flower in wooded places, also the furze, often with an abundance of its yellow blossoms, on heaths and downs, and the hazel with its mature catkins in the hedgerows.

The Chickweed (Fig. 154) and the Red Dead-Nettle (Fig. 156)

The study of the hazel should prove particularly interesting, and we mention this one in particular because its flowers are not so well known and understood as some of the others, except, of course, by those who have made a special study of flowers. The long, drooping catkins will be seen to consist of a number of minute, imperfect (male) flowers, each one consisting of a few little green bracts and four divided stamens that produce abundance of powdery pollen in February or early March. The fertile or female flowers—those which produce the nuts—are grouped together in clusters that are hardly to be distinguished from the ordinary buds of the tree until they protrude their crimson stigmas. This tree affords an interesting example of wind-pollination, the powdery pollen being transferred by the breeze from the stamens of the pendulous catkins to the stigmas of the fertile flowers.

Lastly, we may find further material for the study of the vegetable world, in winter, in the various forms of roots of biennial and perennial plants, and also in the winter stores of plant food laid up in fleshy roots, tubers, corms, rhizomes, bulbs, etc.

C. Animal Life

Seeing that the number of wild creatures available for study during the winter is comparatively small, and that the weather is not generally so favorable for outdoor observations, advantage may be taken of the special opportunity for dealing with the domestic animals, beasts of burden, and the various creatures that may be observed from within doors or in the immediate neighborhood of our dwellings.

In dealing with quadrupeds, as with all other animals, special attention will be called to the movements, particularly those connected with locomotion, procuring the food, and the feeding.

In connection with the last-named process, the teeth must be studied, and the method of using them, including the nature of the movements of the lower jaw. Attention will be called to the characteristics of the biting teeth, and to their position in the front of the mouth, also to the grinding teeth, if such exist, towards the back of the mouth.

It will be noted that carnivorous mammals, such as the cat and the dog, do not masticate their food at all, but that the latter is simply divided into portions that can be swallowed; that the small front teeth (incisors) and, more especially the long pointed canine teeth, are adapted for seizing and tearing the food only, and that the molars at the sides of the jaws simply cut it, with a scissorlike action, into pieces that can be bolted. A prepared skull of one of these animals will be very valuable in demonstrating the uses of the teeth, especially when there is some difficulty in observing their action in the living animal.

Let the children be questioned closely as to the movements they witness. Encourage them to explain, for example, after having seen the working of the molar teeth of the cat, why that animal turns its head on one side when eating a piece of meat. Also, after they have observed the roughness of the cat's tongue, and seen the cat's method of dealing with a meaty bone, get them to explain the use of the rough surface referred to. Again, after they have examined the claws of both the cat and the dog, draw from them the explanation of the fact that, while the cat's claws are always sharp, those of the dog are worn and blunt at the tips. If the children cannot explain these things, the teacher will not give the necessary information, but encourage further observation until the discovery has been made that the claws of the cat are retractile—that they can be withdrawn into sheaths—while those of the dog are not.

The bodies and the corresponding parts of different quadrupeds should always be compared with one another, and the comparisons extended to ourselves. Thus, the children will be led to see that the ear appendages of those animals which have to be always on the alert (at least in the wild state) because of their numerous enemies, or which require a keen discrimination as to sounds in order that they may detect the whereabouts of their prey, are so constructed that they not only collect sound waves very effectually, but also are capable of an independent movement that enables one or both of them to be rapidly turned towards the point from which a sound originates. In our own case, on the contrary, the appendages are not well adapted for the collection and concentration of sound waves, nor are their muscles sufficiently developed to produce any appreciable movement.

It will be noted, also, that the eyes of the more timid herbivorous quadrupeds which, in the wild state, are the prey of carnivorous species, are situated prominently at the sides of the head, thus enabling the animals to see behind them while they are feeding. This is a valuable means towards the preservation of the creatures concerned, since they have to wander out into the open ground for their food, and are thus, during feeding-time, fully exposed to the view of their numerous enemies.

The skillful teacher will always be on the look-out for such examples of adaptation to habit, and for illustrations of features of a protective nature, in his own study of animal life, and will do his best to encourage the children under his care to make such discoveries for themselves.

Not having the space here to point out the leading characteristics of the common mammals individually, we will merely call attention to one other feature of interest—the differences in the limbs and in the mode of using them.

In all cases the limbs of quadrupeds should be compared with those of our own bodies—the front legs with our arms and the hind limbs with our legs; and children should be led to point out in any mammal, those parts and joints corresponding with our own shoulder, upper arm, elbow, lower or forearm, wrist, fingers, and, in the case of the hind leg, the hip, thigh, knee, leg, heel, instep, and toes.

Such studies will enable them to discover that while some animals (called plantigrades, e.g. the rabbit) walk on the soles of their feet, planting the heel on the ground at each step, others, with the heel about halfway up what is commonly regarded as the leg, walk on the tips of the toes of each hind foot, as is the case with the sheep.

Two simple diagrams explaining the general build of a rabbit and a sheep (Fig. 158 & Fig. 159)

Let the children also see that the claws and hoofs of quadrupeds correspond with the nails of our fingers and toes, and that the hoof gives some indication of the number of fingers or toes present. Thus, the single hoof of the foot of the horse is the thickened 'nail' of a single toe, while the divided hoofs of the sheep and the ox are the 'nails' of the two toes present in the foot. They will note, too, that the wrist of the rabbit is practically on the ground, together with the whole of the 'hand,' while the wrists of the sheep and the horse (commonly called the knee) are about halfway between the ground and the lower part of the trunk of the body.

Again, let the children, when examining, for example, the horse, trace the different parts of the legs, both hind and fore, from the foot upwards. They will then find that the elbow and the knee, of the fore and hind-legs respectively, are dose to the trunk of the body, that the thigh bone and the upper arm bone (the humerus) are in the trunk, and that the shoulder and hip joints are quite near the upper part of the trunk. Then, as they observe the horse walking, they will be able to see the movements of the elbow close to the root of the neck, and of the knee, the former with the angle directed backward and the latter with the angle directed forward, as in our own bodies.

The above points are mentioned merely as suggestive examples for the teacher of Nature, and include just a few of the interesting features of mammals which may be observed by the children, under the guidance of the teacher, and which present problems that will afford much food for thought.

As regards birds, winter is undoubtedly a good season in which to entice our feathered friends within easy range for observation. During severe frosts many birds leave the neighboring fields and woods, and closely approach our dwellings in search of food and shelter. Feeding tables, with suitable food, and water renewed very frequently, especially when it freezes, may be the means of bringing them so close that they may be observed through a window at a distance of only a few feet.

In addition to our resident birds—those that remain with us throughout the year—we have now with us the winter visitors, such as the woodcock and the redshank, but most school children are likely to meet with the residents only.

D. Physical Studies

Observations of the rising and setting of the sun, and calculations of the lengths of the days, as suggested for the other seasons, should be continued through the winter months. The children of the lower forms may be taught to measure the length of the shadow of an upright stick cast by the midday sun, and to note how this shadow, gradually increasing in length previous to the end of December, becomes shorter as the season advances; and the elder children, who have been taught how to measure the altitude of the sun, will observe that this altitude remains practically the same for several days at the end of the year, and then gradually increases.

The senior classes will also continue their observations of any of the planets that may be visible at the time, and make simple sketches to show their movements among the stars. The clear, frosty evenings will present very favorable opportunities for mapping some of the more conspicuous constellations of stars; and while this is being done particular attention will be given to some of those groups which are visible only during the present season.

Advantage should be taken of the snowstorm to deal with the structure of the snow crystals and the formation of the snowflake. It will be observed that when, during such a storm, the temperature of the air is below the freezing-point, the snow crystals generally fall singly, while, with an atmospheric temperature above this point, the crystals usually unite to form flakes.

A simple experiment demonstrating that property of ice known as regelation will help to explain the above fact. Break a piece of ice in the schoolroom or in any place where the temperature of the air is a little above the freezing-point, and then put the pieces together in exactly the same position as before: the pieces im-mediately freeze together. Repeat the experiment outdoors at a time when the air temperature is below freezing-point, and the surfaces of the ice, as a consequence, are quite dry, and the pieces will not unite. A film of water is necessary to produce the regelation. Thus, when the snow crystals are quite dry, as they are on very cold days, they will not unite to form flakes, but fall singly.

The pictures, so often given in books, representing snow crystals as seen under the microscope, have apparently led unobservant persons to think that the microscope is absolutely necessary for the study of these beautiful forms. Such, however, is not the case; for the crystals are frequently so large that their forms are clearly discerned without any optical aid. In order to observe them, choose a day when the air is very cold—below the freezing point—and the crystals are falling singly. Allow the snow to fall on a piece of black cloth, and observe them outdoors. Of course a hand lens will enable one to make out the more delicate parts of the crystals more perfectly.

After the above facts concerning snow crystals and flakes have been noted, the teacher should ask the children to explain why the snow is so powdery when the air is very keen; and why they cannot make snowballs so well on such a day as when the air temperature is above the freezing-point and the snow is beginning to melt.

The study of snow crystals will naturally lead to the consideration of crystals in general—their varied forms, the conditions favorable to their formation, and their mode of growth.

Let the children prepare crystals of various kinds for themselves and watch their growth. Some may be prepared by the cooling of a hot saturated solution of some salt, such as alum, nitre, copper sulfate, etc. This method is suitable when it is desired that the rapid growth of the crystals may be observed within a short space of time, but far more perfect forms may be obtained by the slow evaporation, at ordinary temperatures, of saturated cold solutions. If the quantity of solution is rather large, large crystals will be produced, but as the process will then extend over several days, the solution should be protected from dust by means of a covering of fine muslin.

The dry method of producing crystals may also be demonstrated by melting some sulfur, pouring it into a teacup, and, when about half of it has solidified, pouring off the portion still liquid through a hole made in the crust formed at the top.

The crystals prepared by the children themselves, together with any others made by the teacher, and also the crystals easily obtained by other means (such as those of sugar candy) should be observed and sketched by the children. The children will also note that the growth of a crystal differs from that of an animal or plant, the former being due to the addition of like particles added externally, while the latter is the result of new material deposited interstitially.

Outdoor winter studies will include the observation of frost, the crystals of which may be examined with the aid of a lens; and the action of frost in breaking up the soil and in causing porous rocks to crumble; also the nature of icicles, with explanations (given, of course, as far as possible by the children themselves) as to the conditions which determine their form and their vertical position.

A few experiments given by the teacher, dealing with the change of volume that occurs during the solidification of water, will be of great value in explaining the action of frost on rocks and soils, and the bursting of water-pipes in frosty weather. Such experiments will also enable the children to explain why ice floats on water—why the density of ice is less than that of water. The senior children, too, may perform simple experiments with a view of determining the temperature at which water has its maximum density, and then they will be in a better position to explain why water generally freezes at the top first. They may also have the opportunity of observing the presence of ground ice on the bed of a stream, and then of investigating the conditions under which such ice is formed.

Other outdoor observations may include the study of the effects of winter storms and floods, such as the formation of temporary streams, and the effects of such storms on the general appearance of the landscape.

The winter condition of ponds and pools, especially as regards the scarcity of animal and vegetable life, will also be noticed, and, in the case of schools near the coast, the condition of the seashore as compared with that in the summer.

The cold season also suggests a study of the manner in which we attempt to keep ourselves warm. In this connection we may consider the relative values of the different materials used for clothing, demonstrating the meaning of conduction and of radiation by means of simple experiments.

A study of fuels, particularly wood and coal, may also be made, and, in conjunction with this, we may endeavor to make clear the true nature of combustion. Heat-producing foods, and the changes which they undergo within the body, may also be considered; and the process of oxidation, as it takes place in the body, may be compared with the similar chemical actions connected with the combustion of fuels.

Lastly, the formation of coal may be dealt with, the teacher making use of his knowledge in framing a simple account of the history of coal which the children may clearly understand and endeavor to reproduce in their own words.

—8—

Other Studies

The following is a list of other subjects, suitable for school study, that are not associated with any particular season, and which may be taken at any convenient or suitable time, together with hints, here and there, as to the best mode of treatment.

1. Various Animal and Vegetable Products—Their Properties and Uses

Quite a large number of animal and vegetable products provide very useful material for study. In dealing with such it is absolutely necessary that each child has the opportunity of closely examining the material selected, with full permission to handle it and test its principal properties.

We do not recommend the description of manufacturing processes as of any great educational value. Most children will form very vague ideas of such processes from descriptions given by the teacher; and although the descriptions may, to a certain extent, reveal some important relationship between the properties of the material and the manner in which it is prepared or worked, they do not in themselves help the children to observe or think.

A visit to a factory, on the other hand, is highly educational; for it not only gives the children an opportunity of observing for themselves how the different stages of manufacture are carried on, as well as the interesting appliances by which these stages are conducted, but it also gives splendid practical illustrations of the properties of the material which is being worked.

2. The Jetsam of the Seashore

The teachers of schools near the sea will find plenty of material for interesting studies throughout the year, more especially after stormy weather. Let the children ramble along the beach, following the line of miscellaneous debris washed up by the waves at high-water mark. Here they will find numerous interesting seaweeds, some in a very perfect condition, including species that have been detached from submerged rocks and washed ashore by the breakers—species that are seldom or never seen *in situ* down to the level of low tide.

Among the debris will be found various structures of such a nature that the children will be unable to decide as to whether they belong to the animal or the vegetable world, and the investigation of their true nature will afford a most interesting pastime.

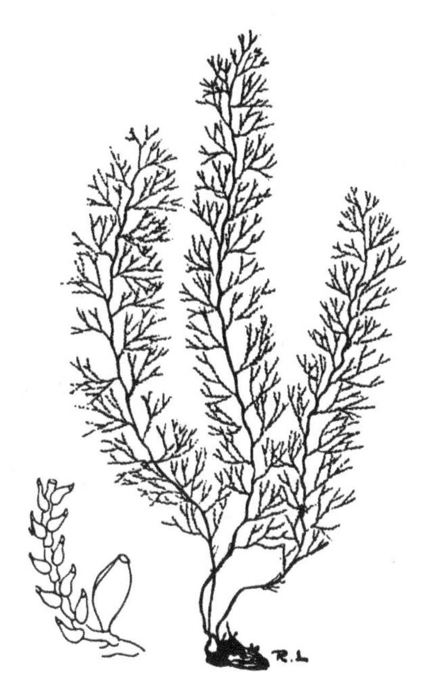

A Sea Fir (Fig. 160)

The sea mat and the sea firs, for example, partake so much of the character of plant forms that they are commonly placed in collections of seaweed; but if a fresh specimen, recently washed up by the waves, be put in a glass of sea-water, and examined by the aid of a lens, the minute animals of the colony, of which the structure is a supporting skeleton, will be seen to protrude from their pores and show signs of active life. On the other hand, the beautiful coralline, which at first might be mistaken for an animal structure allied to certain species of coral, is really a seaweed, the cells of which are supported by rather thick deposits of carbonate of lime.

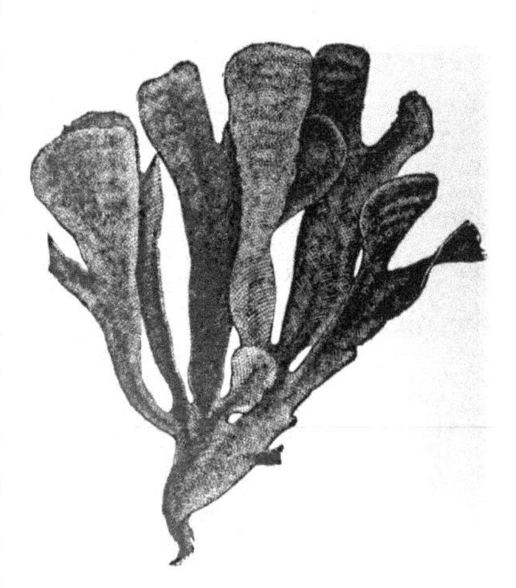

The Sea Mat (Fig. 161)

The miscellaneous matter of the jetsam will also include, among many other interesting things, the cast shells of crabs, the tubes of marine worms, numerous shells of mollusks of the most varied shapes, starfishes, jellyfishes, the egg-cases of skates and dogfishes (and sometimes the complete eggs with the well-formed little fish within), the eggs or spawn of other fishes, the egg-cases of the whelk, the bone (shell) of the cuttle-fish, and various vegetable or other terrestrial matter evidently washed from some other part of the coast, thus throwing some light on the currents prevailing at the time.

3. Various Human Activities

While the children of country districts may always be expected to interest themselves more or less in the different human activities connected with country life and occupations, the town children may also be encouraged to find some connection between certain phases of town life and the outside conditions. Let them always learn to connect the natural or raw productions seen in the town with their proper seasons, and to think out the causes of all changes in the incoming produce and in the corresponding changes in the occupations of the people.

The Coralline Seaweed (Fig. 162)

4. Weather Charts

Even very young children may be taught to take some interest in the changes of the weather, and the results of these changes; but the elder ones will make a much more detailed study of weather changes and climatic conditions. The latter should be led to take regular readings, and to make regular records of, the more common meteorological instruments, such as thermometers, barometer, rain-gauge, etc., and to compare season with season, and one year with another. Let them also learn how the weather chart of the daily paper is made, how it is to be read, and what it teaches. Let them understand the elementary principles on which the daily weather forecasts are based, also that the winds are the prime movers of the weather, and learn the causes which lead to the changes of the wind. Such a training will help them to become very observant, not only of the weather and its frequent changes, but also of their surroundings in general.

5. Rocks and Soils

We have already referred briefly to the observation of the rocks and soils of the locality of the school, but such observations may be carried on at any time, and should receive special notice in cases where any important occupation of the inhabitants is in any way connected with the mineral produce of the neighborhood.

The Common Starfish (Fig. 163)

Let the children study the materials used for building: the bricks (especially if made locally) and the clay from which they are made, the slates or tiles of the roofs, the principal building-stones and paving-stones of the district. The physical properties of these materials should be ascertained by examination on the part of the children, who will be encouraged to state their opinions on the adaptability (or otherwise) of the materials to the uses for which they are employed. As far as possible, these rocks should be examined in situ, and the method of quarrying watched. The disposition of the rock strata should be noticed, taking advantage of the presence of quarries, railway cuttings, etc., for this purpose.

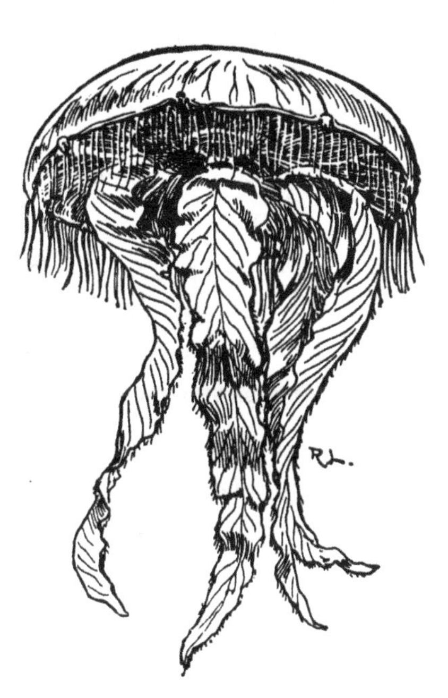

A Jellyfish (Fig. 164)

The children may be asked to account for the disposition of the beds—to explain how the layers were formed. A close examination of the rocks will often give clear proof of the action of water in their formation. They may contain fossil remains that tell of their sedimentary origin, or their particles (as in the case of sandstone, oolite, etc.) may give direct evidence of the action of water. The children will observe all these things, and form their own conclusions, to be corrected when necessary by the teacher, who will supplement the whole matter by telling stories of the past history of the earth as learned from the earth itself, and lead the children to see that changes, similar to those of past ages, are still in progress, and that evidences of this are to be learned in almost every district.

In localities where minerals are mined and worked by the people, the nature of the minerals should be studied, as well as the mining operations and the smelting or other preparation of the metallic or other product from the crude material. This will not only increase the children's powers of observation and create an inquiring habit, but, in many cases, will also directly assist them in the better accomplishment of the work to which they will have to apply themselves in future years.

6. The Forces Molding the Land

Closely allied to the above study is that of the various forces which mold the land. Even the younger children will be able to watch some of the more important denuding agents at work, and to witness the results produced by them in time. The principal of these agents, some of which have been alluded to previously, are wind, rain, frost, running water, and waves. The elder scholars will also get some insight into the nature of the chemical agencies at work to the same end. They will observe how all the above forces tend to reduce the higher ground and to fill up the hollows by transporting material to a lower level; and also witness the results of the gradual accumulation of vegetable matter, especially in marshy and boggy districts, and of the great upheaving forces of the earth, in their counteraction of raising the level of the land.

It will seldom be necessary to take long excursions for the study of the above changes. Much may be learned during a short walk along the banks of a brook, a ramble to a neighboring hill, a stroll around a patch of marshy or boggy ground, or a visit to the nearest cliff by the sea. If the children have the opportunity of observing the above forces at work in a small way, they will readily be made to understand what vast changes may be brought about when the same agencies are active on a much grander scale and for a long period of time. Furthermore, after having observed the nature of clay and sand, the structure of chalk, and specimens of coral and fossiliferous limestone, with the aid of a lens or the compound microscope when necessary, they will be interested in the stories of the formation of clay-beds, sandstones, chalk-beds, limestones, and coral reefs.

7. The Northern Constellations

The Northern (Circumpolar) constellations of stars—those within fifty degrees of the pole star—are visible throughout the year, and may, therefore, be studied at any season. Children should be encouraged to observe these, and to note how they all revolve around the pole star, describing one complete revolution in twenty-four hours, corresponding with an angular velocity of fifteen degrees per hour.

It should be made clear that these stars are fixed, and that the apparent motion is due to the rotation of the earth on its axis.

Attention should first be given to the 'Plough' or 'Charles's Wain,' formed by the seven bright stars in the constellation of the Great Bear. Having become familiar with this group, the attention of the children will be called to the two pointers at the front of the Plough, which give the approximate direction of the pole star. Once familiar with these stars, there will be no difficulty in learning the relative positions and distances of some of the other conspicuous stars and groups of the northern sky.

8. The Milky Way

The form and position of the Milky Way may be observed on any very clear night when the moon is not bright. After the children's attention has been brought to it, and they have been taught the true nature of a star, they are in a position to listen to a simple story of the universe—its vastness and its form.

9. The Magnetic Compass

The properties of the magnet may be studied by the children who are sufficiently advanced to understand the principle and use of the compass. These properties will, of course, be learned from a series of simple experiments, performed by the children rather than by the teacher. Let each child be provided with a magnetized knitting needle, a pinch of iron filings on a sheet of white paper, and a few inches of cotton thread or a saucer of water in which a large cork bung is floating. First they will study the action of the magnet on the iron filings, and observe the apparent distribution of the magnetic force in different parts of the needle. They will then break the needle, and, after testing each part and observing that each is a perfect magnet, will suspend one by the middle on the cotton thread, or float it on the cork, and discover the action of magnets on each other. They will now be in a position to learn, from the directive action of the earth on the suspended or floating magnet, that the earth is itself a magnet.

The children will probably have learned previously how to determine the direction of geographical north by means of the pole star, and of the sun at midday, allowing, in the latter instance, for the variation in solar time; and they will then be in a position to determine approximately the amount of magnetic variation—the difference, in degrees, between the direction of the compass needle and that of the geographical north. Examples will then be given of the use of the compass both on land and at sea, with the mode of correcting for the true geographical direction.

— 9 —

Seasonal Nature Study Course Outline

The following outline, arranged for the convenience of the teacher when making out his scheme of work, includes the majority of the subjects referred to in the preceding chapters.

It would be impossible to deal thoroughly with all the subjects mentioned below during the period of a child's school life; and it must be understood that we do not propose the whole as a scheme for school study, but rather as a list of subjects from which a selection may be made according to the inclinations of the teacher and with reference to the nature of the neighborhood of the school.

It will be observed that no attempt has been made to divide the whole outline into separate schemes of work for children of different forms or ages. This, we take it, is quite unnecessary. The teacher will be competent to make his own selection of subjects, aided by the suggestions given here, and also to adapt his methods to the capacities of the children under his care.

Although there are certain subjects peculiarly suitable for the training of young minds, and others which are better adapted for treatment with senior scholars, yet we are of opinion that by far the greater number of natural objects and natural phenomena are useful studies for school children of all ages.

This being so, it appears to us to be a great mistake to encourage young children to observe a certain object, and then to make no further reference to that object during the remainder of their school life.

In the lower forms the children will observe, think about, and describe only the more obvious characteristics of a natural object. A year or two later, their powers of observation and reasoning having considerably developed, they will look upon the same object with different eyes, see a great deal more than they saw before, and think over problems that were formerly quite beyond their grasp. Hence it is very profitable to arrange the nature study scheme in such a manner that some of the subjects, at least, appear two or three times, to be dealt with during as many different periods of a child's school career.

Thus, when very young children observe a living fish, they note its general form, but without the power to grasp the adaptability of the form to the movements of the fish in the water; they observe the fins, and even note their motions, but without the possibility of discovering and understanding functions of individual fins in propelling, steering, and balancing the body; they see the movements of the mouth and of the gill-covers, while as yet they are not in a position to understand the use of these movements and the manner in which the fish breathes. If, however, a child is encouraged to study the living fish two or three times during its school career, say at intervals of one or two years, each lesson will be quite as useful, and, perhaps, more so, than three lessons on separate subjects taken at the same periods, for in each case the child assimilates just that which is adapted to the present state of its growing mind.

We leave the teacher, then, to use his own judgment in the selection of suitable subjects from the outline provided, feeling confident that the actual experience of one who is in real earnest as regards the training of the young under his care will serve him better than any hard and fast rules, especially if that experience is matured by a communion with others whose labors are directed to the same end, and by a careful study of the development of the child mind.

A. Spring Studies

Spring the season of the re-awakening of life.

Vegetable Life

The opening of buds.

Detailed study of one large bud:

Bud-scales and other temporary structures.

Gradual transition of bud-scales into leaves.

Simple experiments to demonstrate the manner in which the sap flows.

Germination of various seeds under different conditions as to moisture, food, heat and light. Records kept.

Plants reared from seeds, in a good soil, for continuous observation. Records of life-history.

The growth of bulbs and corms.

The growth of potato plants from the tubers under varying conditions. Make records.

Spring flowers (chiefly outdoor studies).

Habitats and habits.

Calendar of observations.

Cultivation of flowers in the school garden.

Animal Life

Forms and habits of the common creatures of the garden—snails, slugs, centipedes, young spiders, etc.

Rearing of caterpillars or other insect grubs for the study of their metamorphosis.

Observations of aquatic creatures in the school aquarium:

Development of Frog eggs.

Various aquatic larvae.

Water snails. Small fishes.

Marine life as seen in the rock-pools.

Study of the common birds of the neighborhood:

Return of the summer visitors.

Nest-building and the care of the young.

The common mammals of the neighborhood:

Wild and domestic. Forms and habits.

The frolicking of young animals—lambs, kittens, etc.

Studies of Earth, Air and Sky

Daily path of the sun: rising, setting, altitude at midday.

Lengthening day and increasing warmth.

Spring winds and showers. Droughts and dust.

Planets visible at the time. Appearances and movements.

Stars. Their apparent motions. Conspicuous constellations.

B. Summer Studies

Summer the season of the greatest abundance of animal and vegetable life, and of rapid growth and development.

Plant Life

Summer wildflowers—chiefly outdoor work:

Observations of habitats and habits.

The flowers and weeds of the garden:

The struggle for existence.

How plants are protected—thorns, spines, prickles, etc.

Forms and arrangement of leaves. Leaf mosaics. Functions of leaves.

Storage of food in rootstocks, tubers, bulbs, etc.

Calendar of summer flowers. Records of observations on the habitats, habits, flowering, fruiting, etc.

Parts of the flower and their uses:

Relation between flowers and insects.

Relationships in plants, as shown in the structure of the flowers and other parts.

Our forest trees and shrubs. General form, bark, branching, leaves, fruit, etc.

Simple experiments illustrating the general activities of plants:

Absorption of water, transpiration, movements of sap, formation of starch and other products.

Flowerless plants and their life histories:

Ferns, mosses, lichens, fungi, algae.

Animal Life

The small creatures of the garden.

Common birds of the neighborhood.

Habits of animals seen during school rambles.

Common creatures of our ponds and streams.

Life in the rock-pools on the coast.

In all the above attention paid particularly to—

Movements—voluntary and instinctive.

Means of defense and offense.

Means of capturing or procuring food.

Manner in which the food is eaten.

Construction of homes or shelters.

Solitary and social life.

Storing of food not required for immediate use.

Care of the young: preparations for, protection, feeding, teaching.

Construction of snares—spiders.

Resemblances to environment, and mimicry

Studies of Earth, Air and Sky

The sun: rising, setting, altitude at midday.

Length of the day. Summer temperatures.

Summer showers and droughts. Their effects.

The planets visible.

The star-constellations visible in summer only.

C. Autumn Studies

Gradual reduction in temperature and gradual decline in both animal and vegetable life.

Vegetable Life

Ripening fruits. How fruits are formed.
Difference between fruits and seeds.
Uses of the fruits (seedcases) to the seeds within them.
Splitting and non-splitting fruits.
Collection of fruits for study and classification.
Distribution of fruits and seeds. Agents concerned.
Autumn flowers—studied, as far as possible, in their habitats.
Decay of leaves. Autumn tints.
Fall of the leaf. Cause of. Observations and records.
The meaning of decay. Action of bacteria.
Storage of food by biennials and perennials.

Animal Life

Creatures that never live to the end of the year:
Deposit of eggs before they die.
Storage of food for the winter—squirrels, bees, etc.
The movements of birds. Summer visitors leaving. Winter visitors arriving. Birds of passage.
Small creatures of the garden seeking shelter for the coming winter.

The Earth, Air and Sky

The shortening day and decreasing temperature.
Observations of the rising and setting sun:
Decreasing altitude of the midday sun.
Autumn gales, mists, and fogs.
The planets visible at the time.
Some constellations of stars visible only during the autumn.

D. Winter Studies

Winter Studies Life now at its lowest ebb. Many plants and animals in a dormant condition.

Plant Life.

Winter condition of deciduous trees and shrubs.
Study of winter buds.
Evergreens: their principal characteristics.
Conifers and their cones.
Winter condition of hedge, field, wood, and moor.
Winter flowers: snowdrop, hazel, furze, etc.
Early flowers in sheltered places.
Winter condition of herbaceous biennials and perennials: their roots, tubers, corms and bulbs.

Animal Life.

Hibernating animals: their homes and their condition.
Winter hiders, including the little creatures of our gardens.
Dormant stages (pupae) of insects.
Birds seen in winter.
Animals that change their covering for the winter:
Advantages of the change.
Winter life of the squirrel.
Domestic animals.
The care of flocks and herds in winter.
Bees in winter.
Queen humble-bees and queen wasps.

The Earth, Air and Sky.

The snowstorm. Snow crystals and flakes. Compare with other crystals.
Frost and its action.
Ice and its properties. Icicles.
Winter storms and floods.

Winter landscapes. Contrast with summer.
Winter condition of ponds and pools.
The seashore in winter.

E. Other Studies

Other Studies (For any time of the year)

Vegetable products: their properties and uses.
Animal products: properties and uses.
The jetsam of the seashore.
Various human activities in town and country.
Weather charts: how made, and their use.
The rocks and soils of the neighborhood:
  Building and paving stones. Their properties.
Other mineral products of the neighborhood.
Disposition of rock-beds in the locality.
The forces molding the land.
  Streams and their action.
  Action of the sea on the land.
  The atmosphere as a denuding agent.
Clay, chalk, coral, and other interesting rock-formations.
The magnetic compass: its principle and use:
  How to find the geographical North by means of the compass, the pole star, and the sun.
The northern constellations of stars always visible:
  Their apparent daily motion.
The Milky Way. The universe.

— 10 —

Classification of Animals

A knowledge of the classification of animals—of the principal groups into which the various members of the animal kingdom have been placed, and the leading characteristics common to the animals in each group—is very helpful to a teacher; for although he may never deal with the principles of classification as a subject of study for his class (and this subject should not be chosen except for senior classes which have previously observed a considerable variety of animal forms), yet a knowledge of these principles will assist him in directing the children's attention to the most important of the obvious characteristics of the creatures under observation, and to correct those many confusions that so frequently arise from the inappropriate and misleading popular names by which those creatures are often known.

For example, even young children may be led to learn, by their own observations, that the bat is not a bird because it flies, that whales and seals are not to be regarded as fishes merely because of their aquatic habitats, that the snake-like eel differs considerably from snakes in structure and habit, and that the term insect is not to be applied promiscuously to all little creeping things. On the contrary, the teacher will invariably call attention to those leading features of each animal observed as will enable the children, in later years, to associate it with the others of its class, and help them to form some estimate of the position it holds in the scale of animal life.

Such careful teaching will also prevent the children from being misled, as we have already hinted, by more or less inappropriate popular names. Thus, they will be led to see that the interesting little blindworm, as well as the glow-worm, silkworm, mealworm and bloodworm, are not worms, but that they differ in important respects from the members of this latter class. Also that shellfish, starfishes, cuttlefishes, and jellyfishes are quite distinct from the true fishes.

Although the teacher would not find it necessary to call the attention of the children to the class-distinctions, as such, of the different animals observed, except, perhaps, in the upper forms, yet he himself should possess such an elementary knowledge of these class-distinctions as will enable him to lead the children to note the important points of structure, and thus put them in a position to form their own ideas, more or less accurate, of the relationships of the creatures they have seen.

It is with this object that we give the following summary of the principal divisions of the animal kingdom. And though, as it will be seen, we call attention almost exclusively to those obvious characteristics which may be observed by the youngest scholars, yet occasionally we refer to features outside the range of a child's observation, but permissible in the case of the elder children who have acquired a rudimentary knowledge of anatomy and physiology.

VERTEBRATE ANIMALS

Animals with internal bony (in some cases cartilaginous or gristly) skeletons, the chief part of which is the backbone or vertebral column. The body contains two tubes or cavities, one behind (or above) the bony axis, containing the chief part (brain and spinal cord) of the nervous system, and the other in front of (or below) the bony axis, containing the organs of digestion and circulation. This division of the animal kingdom is sub-divided into:

1. Mammals.

The highest of the vertebrates. The young of mammals are born in a living state, and are suckled by the mother. The body is usually more or less covered with hair (or some modification of hair). They have warm blood, and a heart with four divisions, and they breathe by means of lungs, nearly all of them being terrestrial in their habits. This group includes mankind, monkeys, bats, all quadrupeds or four-footed beasts, seals, and whales.

2. Birds

Birds are clothed by a covering of feathers, and the forelimbs are modified into wings for flying. They have warm blood, a heart with four divisions, and they breathe by lungs. Birds have no teeth. Their young are produced from eggs.

3. Reptiles

Reptiles have a covering of scales. Their blood is cold, and the heart has only three divisions (two auricles and one ventricle). They breathe by means of lungs. The young are generally produced from eggs. The group includes lizards (Fig. 71), tortoises, turtles, crocodiles, and snakes.

4. Amphibians

These animals have a smooth (generally), naked skin, and begin life as little, fishlike, limbless 'tadpoles' which live in water and breathe by gills. At a later stage the gills disappear, lungs are formed, limbs are developed, and the animals live on land. The amphibians include frogs, toads, and newts (Figs. 68 and 69). They have cold blood, and a heart with three divisions.

5. Fishes

Fishes are aquatic in habit, and are peculiarly adapted for a life in water. The body is covered with scales, and the four limbs are developed into fanlike fins. Fishes have cold blood, and a heart with only two divisions, and they breathe by gills. The young are produced from eggs.

INVERTEBRATE ANIMALS

The invertebrates have no internal bony skeleton, but their bodies are often protected and supported by a calcareous shell or a hardened skin. The chief divisions are:

1. Crustaceans

The bodies of crustaceans consist of a number of segments surrounded by a horny or calcareous skin. Between these segments the skin is usually unhardened, so that the former are movable. The head and thorax are usually welded into one, and the former is provided with antennas and compound eyes. The legs, generally numbering five pairs or more, are jointed. Crustaceans are mostly aquatic, breathing by gills or absorbing oxygen through the skin. The young are produced from eggs, and undergo metamorphosis while still very young. This division of the invertebrates includes crabs, lobsters, crayfishes, shrimps, prawns, sandhoppers, and woodlice, besides many microscopic creatures.

The Common Shore Crab (Fig. 166)

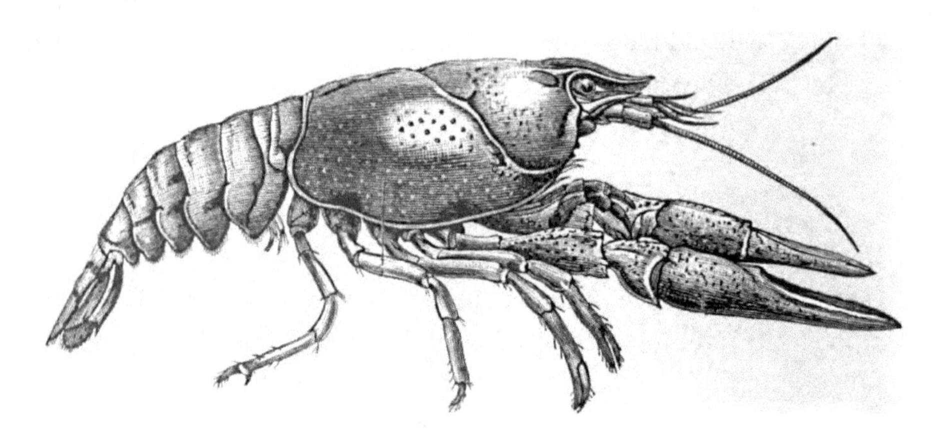

The River Crayfish (Fig. 167)

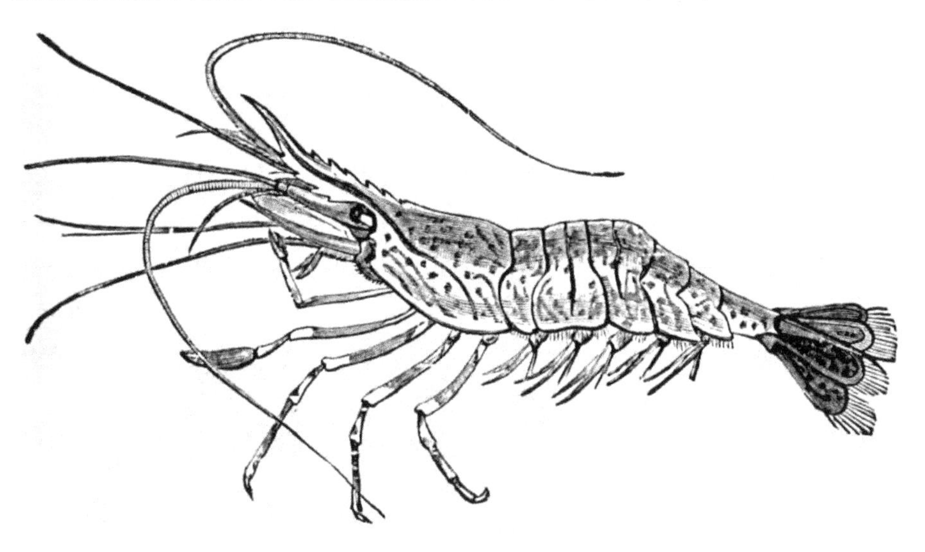

The Prawn (Fig. 168)

2. Insects

The bodies of insects consist of three more or less distinct parts—the head, thorax, and abdomen. The head bears a pair of antenna or feelers, and a pair of compound eyes (eyes divided into a number of compartments, each with its own lens). The thorax is made up of three segments, each of which bears a pair of legs, and this portion of the body also generally bears one or two pairs of wings.

The abdomen is made up of several ring-like segments. Most insects undergo remarkable changes in form (Fig. 41), appearing first as a larva or grub, then as a pupa or chrysalis, and finally in the winged or perfect state. The characteristics of the insect given above are those of the perfect or final stage.

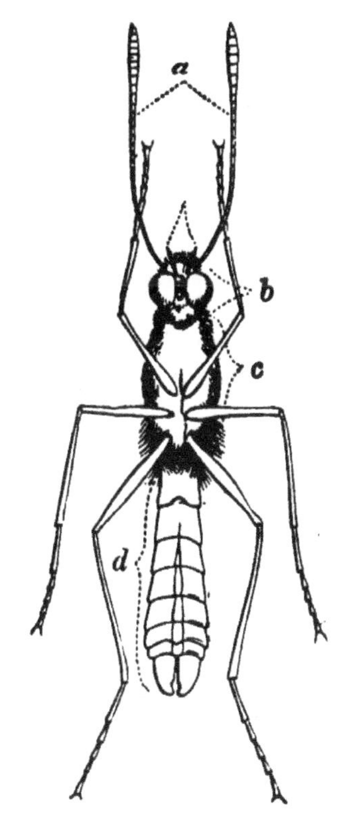

The body of an insect (Fig. 165)

a. antenna, b. head

c. thorax, d. abdomen

3. The Spider Class (Arachnoidea)

The creatures of this class are terrestrial, and air-breathers. The head and thorax are united, the former with a pair of jaws connected with poison glands, and simple eyes, often six or eight in number; and the latter with four pairs of jointed legs. Spiders have silk-spinning organs (spinnerets) at the tip of the abdomen. This class contains spiders, scorpions and mites.

The Jumping Spider (Fig. 169)

The Sandhopper (Fig. 170)

4. The Many-legged animals (Myriapoda)

In this division the body is long, narrow, and made up of many similar segments each of which bears a pair of jointed legs. The head has a pair of antennae and several simple eyes. The animals of this group, which includes centipedes and millipedes (Figs. 39 and 40), are mostly carnivorous, and nocturnal in their habits.

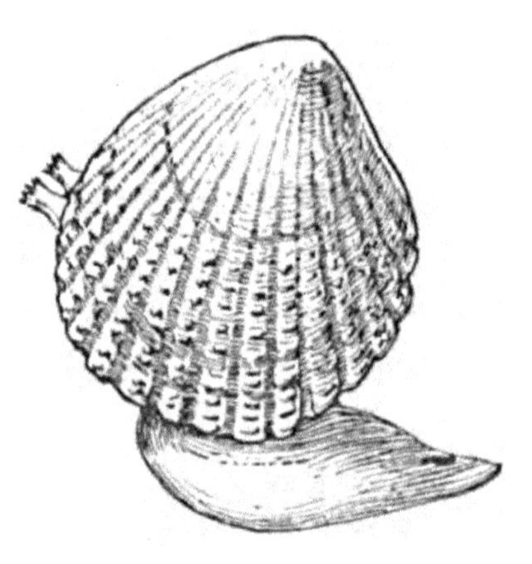

The Cockle (a bivalve) with its fleshy 'foot' extended (Fig. 171)

Note. *The above four divisions of the invertebrates are usually grouped together and called the Jointed Animals (Arthropoda). They are all characterized by their segmented bodies and jointed limbs.*

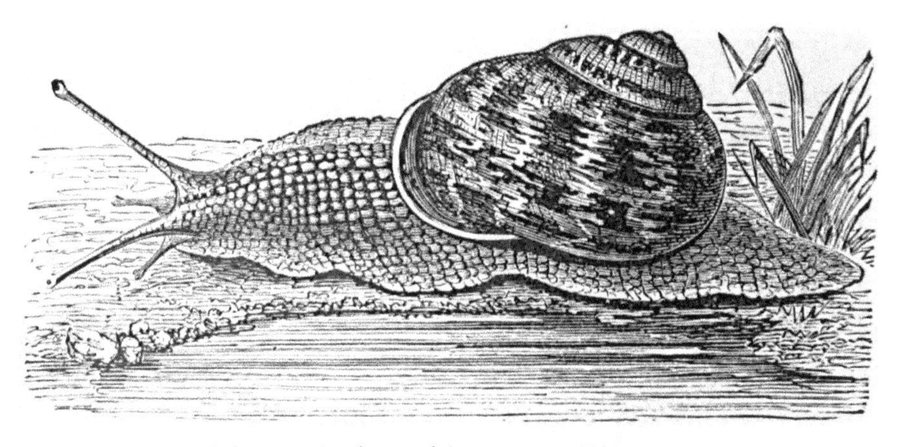

A Common Snail, one of the gasteropods (Fig. 172)

5. Mollusks

The mollusks are soft-bodied animals, not segmented, without limbs, and usually covered with a leathery mantle which secretes a shell of calcium carbonate. This group is sub-divided into:

a) *Headless Mollusks or Bivalves*, with no distinct head. They are enclosed in a shell of two parts, and are aquatic, breathing by gills. Examples: mussel, oyster, cockle, and scallop.

b) *Headed Mollusks or Gasteropoda*, with a distinct head, and shell (if present) in one part. The animals of this division creep on the undersurface of the body. Some are aquatic, breathing by gills, and others are terrestrial, breathing by a lung. Examples: snail, slug, winkle, whelk, and limpet.

c) *Head-footed Mollusks*. These have a distinct head, with two large eyes and with tentacles bearing suckers. Most of them are not enveloped in shells, but they have generally a shell that is enclosed within the body. Examples: cuttle fish, squid, calamary, and octopus.

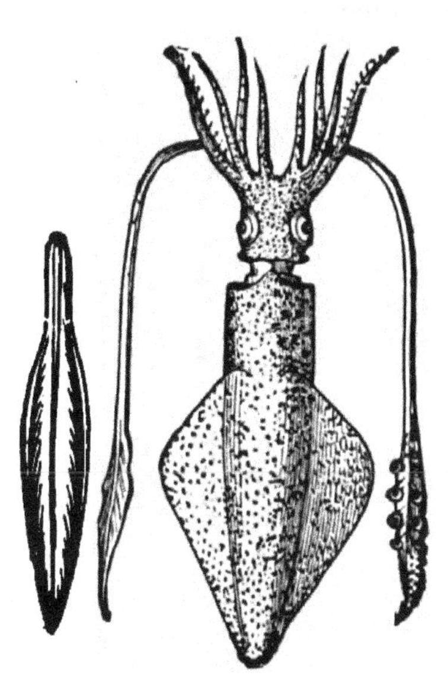

The Calamary (a head-footed mollusk) and its shell (Fig. 173)

6. Worms

Animals with soft, elongated bodies, generally made up of many ring-like segments, and no limbs. Worms have usually no hardened covering, but some marine species live in tubes constructed with sand, while others dwell in calcareous tubes secreted by the body.

The Lugworm, common on our sandy shores (Fig. 174)

7. Spiny-skinned Animals (Echinoderms)

These animals have a spiny skin, and the parts of their bodies are regularly disposed around a common center. They move about by means of little tubular feet which are filled with water from within, and which terminate in sucking discs. They include starfishes, sea-urchins, and the sea-cucumbers.

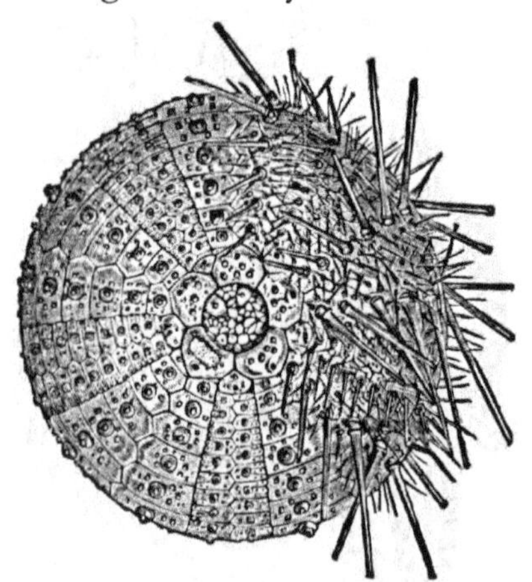

Sea Urchin, spines removed (Fig. 175)

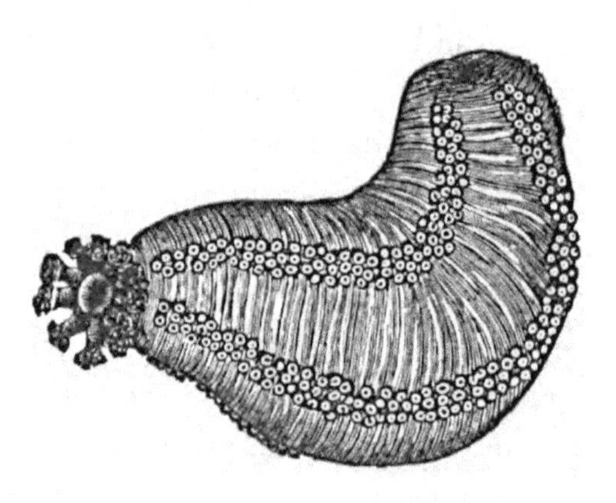

Sea Cucumber (Fig. 176)

8. Jellyfishes

The jellyfishes have soft, jelly-like bodies, circular or cylindrical in form, with a number of tentacles arranged around the mouth. They are very simply formed, having no internal organs except a stomach, and no organs of sense. The tentacles are armed with stinging cells by means of which the animals can paralyze their prey. Jellyfishes are reproduced by means of eggs, and also by means of bud-like growths which sometimes remain attached to the parent animal, but are often detached and lead a separate existence. This division includes the beautiful sea-anemones of our rock-pools (Fig. 37), the floating jellyfishes that abound in our seas (Fig. 164), and the coral-building jellyfishes, which have calcareous skeletons. Also the little creatures which live in colonies, supported by a common skeleton that is popularly known as the sea-fir, so commonly seen on our shores (Fig. 160).

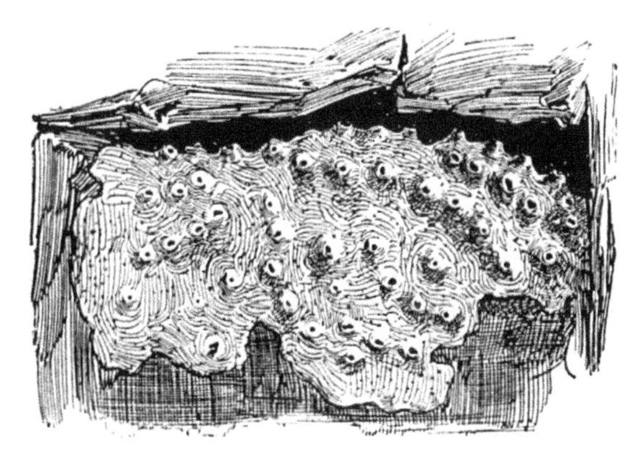

A common British sponge (Fig. 177)

9. Sponges

These consist of a number of microscopic cells, all united into one mass, and usually supported by a skeleton of calcium carbonate, silica, or horny material. Each sponge has one or more large holes, besides many small ones, and water is kept continuously circulating through the system, entering through the small holes and escaping through the larger ones, the currents bringing constant supplies of air (dissolved) and food. Sponges are reproduced by eggs, and also by division. If a living sponge is cut into pieces each piece will develop into a new sponge.

Many sponges of various forms are to be found attached to the rocks and weeds of our shores.

10. Protozoa

The protozoans are microscopic animals, each consisting of a single cell, and often displaying no differentiation in structure. They are mere specks of living protoplasm (a jelly-like substance that occurs in all living beings) capable of performing all the functions which, in higher animals, are performed by special parts or organs. Some of the protozoans partake somewhat of the nature of vegetable cells, and allied to them are minute organisms which constitute the borderland between the animal and the vegetable kingdoms, belonging neither to the one nor to the other. Protozoans are to be found in abundance in water, on soil—in fact, everywhere.

— 11 —

Classification of Plants

For reasons similar to those given in the previous chapter we are providing an outline of the classification of vegetable life.

FLOWERING PLANTS (PHANEROGAMS)

Plants producing seeds, each of which consists of, or contains, an embryo plant with seed leaves (cotyledons), bud (plumule), and young root (radicle), and which separates from the parent plant on reaching maturity.

1. Angiosperms

Plants in which the seeds (ripened ovules of the flower) are enclosed in a fruit (the ripened ovary).

a) *Dicotyledons*: Plants the seeds of which have two cotyledons or seed leaves. They are generally characterized by a stem consisting of a central pith, around which is wood arranged in a ring or in rings, and an outer epidermis or bark. The veins of their leaves usually form a network, and the parts of the flower are frequently arranged in whorls of four or five or multiples of four or five. This division contains by far the larger number of our flowering plants.

b) *Monocotyledons*: Plants with only one seed-leaf in the seed. The stem has no pith, and no true bark, and the wood is not arranged in rings. The leaves usually have parallel veins, without any network; and the parts of the flower are arranged in whorls of three or a multiple of three. This group contains lilies, daffodils, orchids, grasses, etc.

2. Gymnosperms

Plants in which the ovule is not enclosed in an ovary, but either in the axil of a carpel or attached to the surface of a carpel. The division includes the cone-bearing trees, such as pines, firs, larch, yew, etc. In most cases the cones are formed of woody scales (the matured carpels of the flower) on which the naked seeds lie.

FLOWERLESS PLANTS (CRYPTOGAMS)

These plants are not produced from seeds, but from minute cells, generally called spores, which do not contain the parts of the first produce a cellular thread, a mass of such threads, a cellular membrane, or a cellular, leaf-like body, which afterwards gives rise to the fruit or to a plant producing fruit.

1. The Vascular Cryptogams

These plants have true roots, stems and leaves; and some of their cells are modified into vessels which are arranged in distinct bundles. Their spores first produce a leaf-like body (the prothallus), and from this the future plant arises.

a) *The Club-moss Group*: In this group the stems and the leaves are very variable, but the spore-cases are always produced in the axils of the latter.

b) *Horsetails*: In these the leaves are small, and arranged in whorls around the stem; and the upper leaves, which bear the spores, form a spike at the top of the stem. Most of the horsetails grow in marshy or wet places.

c) *Ferns*: The ferns have well-formed leaves or fronds, distributed on all parts of the stem; and the spore-cases are produced on the undersurface of the leaves.

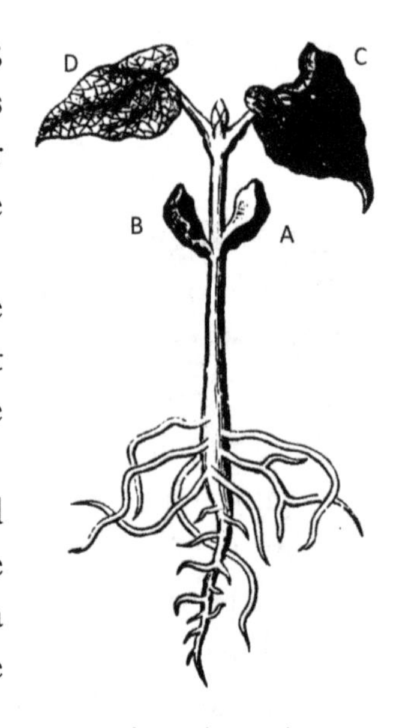

A young bean plant with its two cotyledons (a & b) and its first foliage leaves (c & d)(Fig. 178)

2. The Moss and Liverwort Group

The plants of this group have stems and leaves, but no true roots; they are also entirely cellular in structure, without any bundles of vessels as in the last division. The development of the plant is also different, for the spores give rise directly to the plant, and this latter then produces a capsule in which spores are contained.

A fern (Fig. 179) *Moss Plant (Fig. 180)*

a) *Mosses*: The mosses have both stem and leaves; and their capsules open, when ripe, by means of a lid.

b) *Liverworts*: Some of these plants have a distinct stem with leaves, but they generally consist of leaf-like bodies with root hairs on the undersurface. The leaves consist of a single layer of cells, and the capsule has not a distinct lid.

3. The Leafless Plants (Thallogens)

These have no root, stem or leaf, but the plant consists only of a cellular body with, sometimes, leaf-like expansions.

a) *Fungi the Mushroom:* The fungi have no chlorophyll (the green coloring matter of plants), and consequently they cannot build up their organic material from mineral substances, but must have organic food. They increase by means of spores or by division. The group includes yeast and other unicellular, microscopic plants, molds, mildews, mushrooms and 'toadstools.'

b) *Algos*: The algae contain chlorophyll, and can build up organic material from inorganic matter. Some are unicellular and microscopic, while others are of immense size, and are variously colored—green, olive, brown, red, etc. They multiply by spores or by division. The algae include seaweeds, diatoms, desmids, etc.

NOTE: *Lichens, so commonly seen clothing rocks and the branches of trees, consist of fungi that are parasitic on algos.*

A piece of dead branch with Lichen (Fig. 181)

— 12 —

The School Museum

A. What the Museum Should Contain

Although the school nature lessons should be almost invariably the study of living things in their natural surroundings, of fresh specimens procured for closer study at home or in the school, or of natural phenomena as they occur, yet there are times when it is useful to refer to things out of season, especially when it is desired to compare and contrast objects in their season with the same as obtained at some other period of the year; and for this, as well as for other purposes to be stated presently, it is exceedingly useful to have a museum of natural objects and records for reference as occasion may require.

The teacher will do well to allow, and even to encourage, the children to take a real interest in any collection of material prepared for such a purpose. Let them, under proper guidance, assist in the work of collecting useful things, as well as in the preparation, preservation, arrangement and labeling of the specimens required.

As regards the collecting, however, the teacher's instructions to the children should be somewhat definite, or much useless material will be brought in; and if such useless material is rejected by the teacher, the children, who have but little idea of the relative value of objects for educational purposes, will naturally be discouraged and lose interest; while, on the other hand, if all material procured be accepted and stored, in order to give encouragement, the museum will soon become charged with useless objects that will never be observed.

The vast number and variety of natural objects that may be considered of value in a school museum renders it impossible to provide an exhaustive, or even satisfactory, list for general guidance; but we propose to make some few suggestions as to what the museum may contain, leaving, for the present, all reference to matters connected with collecting, preserving and labeling.

First, then, as regards vegetable life, a collection of the dry fruits of our trees and flowering plants will be exceedingly useful for reference, especially if they are properly mounted, and classified according to their modes of dispersal, and, in the case of dehiscent fruits, according to the manner in which they split. Although these fruits will all have been studied in their season, yet it will be interesting to refer to them at times when the leaves of their plants are being observed, and to associate them with the flowers that produce them.

Collections of leaves will also prove useful at times, particularly if such collections have been prepared with some definite object in view—thus, sets of leaves, each set mounted on a separate card, to illustrate (1) simple leaves, (2) compound leaves, (3) the transitions from simple to compound, (4) autumn tints, etc.

As regards the transition referred to, very interesting examples may be obtained from the bramble and from the Virginia creeper (*Ampelopsis*), on both of which various intermediate forms between the simple and the compound occur. And well-preserved specimens illustrating autumn tints are not only objects of beauty, but they will be useful for comparison with leaves of the same species in spring and summer.

Specimens of the wood of our principal forest trees will prove useful when endeavoring to understand the mode of growth. Such specimens should display transverse, longitudinal, and tangential sections, to show clearly the nature and arrangement of the rings; and sections made through points where branches originated will serve to explain the true nature of a knot.

Dried plants and flowers are not to be recommended. They are never satisfied for teaching purposes. In fact, for ordinary schoolwork they are absolutely useless. As we have previously stated, plants and their flowers should always be studied alive and, whenever possible, in their habitats.

Collections of seeds are also of little value. Seeds should be procured in their season, and preserved only until they are required for the study of their structure and germination. School museums commonly contain large numbers of bottles filled with grain and seeds of various kinds. These are seldom, if ever, needed for reference, and most of them can be obtained very readily whenever they may be required. It is a great mistake, in any case, to provide space for objects that may be procured at any time, such as the produce of the grocer's and the corn-chandler's stores.

Speaking generally, there is not much to be gained, from the nature-study point of view, from an extensive collection of vegetable products and specimens, since the work is, or should be, the study of living things in their proper seasons. In connection with the teaching of geography, however, such a collection may prove more useful, especially as illustrating the products of other countries.

Concerning the animal world, it is probable that the study of this branch of Nature, if taken up somewhat fully, will make greater demands on the museum space. We will give some few examples of specimens which we regard as useful in a collection prepared for the purpose of reference.

Whilst studying live mammals, and observing the manner in which they dispose of their food, it is necessary to refer to the nature of the teeth, their arrangement, and their functions. But, at least in many cases, it is hardly possible to observe the teeth in the live animal, and thus it is well to have specimens of the different kinds. A few skulls of common mammals, with all the teeth in situ, will be further useful in demonstrating the manner in which the teeth perform their functions. For instance, the skull of a carnivorous animal will show perfectly the scissors-like action of the cutting molars, while that of the sheep will serve to demonstrate the mill-stone action of the grinding molars of the herbivorous species.

Where the nature study is so extended as to include some elementary notions of the anatomy and physiology of the human frame as introductory to, or explanatory of, the theory of physical exercises or the elementary principles of hygiene, various specimens and models will afford much help.

Cleaned bones obtained from the rabbit, or from a shoulder or leg of mutton, will give splendid opportunities of observing the structure and action of various kinds of joints. A few consecutive bones from the vertebral column of the rabbit will suffice to give a good idea of the nature of the backbone. The composition of bone may be well illustrated by means of two specimens, similar in size and form, one of which has been soaked in dilute acid to remove all its mineral matter, and then preserved in fluid (see page 196), while the other has been burnt in order to destroy its animal or organic portion. Many useful things may also be selected from the kitchen waste material and profitably employed in the school, and among such we may mention the trachea (or a portion thereof) of any mammal used as food; the larynx (cleaned and preserved) of the sheep, obtained with the sheep's head as supplied by the butcher, etc.

Bird nests and eggs are commonly included in school museums, and it is probable that the acquisition of these is generally the result of a plunder that should on no account be encouraged. If nests are preserved at all (and some of them are really wonderful objects), they should only be those which have been discarded by the birds that have reared their young in them; but it should always be remembered that a great deal of the interest attached to the bird's nest is destroyed as soon as it has been removed from its natural position. The observation of bird nests should be made outdoors, where the positions and the artifices for concealment may be witnessed. Then, as regards the eggs, they should be studied while in the nest and not removed for the purpose of making a collection.

The interest attached to them is to a great extent gone when they are viewed apart from their surroundings, but if it is desired to procure a few specimens, only those should be taken that have been deserted by the parents. In some few instances the eggs of birds possess some special feature of interest beyond the color and markings. That of the guillemot, for example, is of such a form that it will not easily roll off the naked ledge of rock on which it is laid. When eggs are required for the purpose of illustrating such special features, they should, as far as possible, be those of our food-birds, obtained from the poulterer, rather than those procured by the plunder of the nests of other species.

While speaking of birds, we may refer to the wonderful variety of form and color exhibited by their feathers. A very instructive collection of the feathers of our food-birds may be obtained from a poulterer, and typical kinds may be mounted on cardboard for reference. The study of a feather is particularly instructive, and the school museum might well contain some sets of different kinds, sufficient of each set for distribution to a class as required.

This remark reminds us of a second use of the school museum to which we have not previously referred. In addition to the specimens displayed for general reference, the museum may include a store department in which material is kept for distribution to the children when it is desired that they shall all make a study of the same object at the same time. Many interesting natural objects are so common, and occupy so little space, that they are very conveniently stored for the purpose we mention. We shall presently refer again to this subject, and provide a list of material suitable for the purpose mentioned.

We have already made it clear that the study of animals should be made from living specimens, and, in the case of wild species, in their habitats—that apart from the interesting movements the study of animals is robbed of the greater part of its value. Stuffed mammals and birds are quite out of place in a school museum. There are sufficient domestic and wild species to occupy the children's attention without the introduction of dead material; and when, apart from nature study, it is desired to give information on certain other forms, good colored pictures, it seems to us, are often preferable to the dry, rigid, glass-eyed specimens from the taxidermist's stores. The former are equally pleasing in appearance as a well-stuffed specimen, and possess advantages inasmuch as they are more easily stored, are not attacked by moth and other museum vermin, and can be so artistically represented in their natural surroundings as to produce really beautiful pictures.

There are times, however, when it is useful, and even advisable, to have preserved specimens of various forms of animal life for reference.

For instance, during school rambles we may now and again meet with some uncommon creature of special interest. This, after having been kept alive as long as possible for the observation of its habits, may be preserved in spirit or other preservative for future reference.

Again, there are many interesting creatures that are not easily observed alive, either on account of their small size, their retiring habits, or because of their venomous nature. The small creatures referred to may be preserved in specimen tubes in order that they may be closely examined by the aid of a lens, and the same remark, as to preservation, applies to the venomous species.

All country children at least should be taught to observe the distinguishing features of the two common British snakes—the viper and the grass-snake—and of the snake-like lizard known as the slow-worm or blindworm, in order that they may not be fearful of, or led to destroy, pretty and harmless creatures under the impression that they are venomous. The only venomous British reptile is the viper; and since this one should not be kept alive for observation in the schoolroom, unless under the superintendence of a teacher who is thoroughly acquainted with the management of such creatures, and provided with perfectly safe accommodation, a preserved specimen will be useful for the purpose of acquainting the children with its distinguishing markings.

As a general rule, preserved animals should not take the place of living ones for study, nor should life ever be taken in order to provide museum specimens, unless such specimens are absolutely necessary for certain special purposes for which the living animal is not suitable.

Insects, especially the more brightly-colored butterflies and moths, are often killed in large numbers in order to produce showy cases for the museum; and the same creatures are often reared for the purpose of making up sets to illustrate life histories. Such cases and sets are often very beautiful and not without interest, but the exhibition of them does much, we fear, to encourage the wanton destruction of life, and to discourage the sympathetic observation of living things.

And there is no real need for the possession of specimens such as we have described, since quite a large number of insects, including some of our most beautiful species, are so easily reared in captivity that their whole life histories and habits may be perfectly observed.

The various kinds of coverings of animals and the homes or nests constructed by them are always useful in the school museum. The coverings include many modifications of outgrowths of the skin, as well as protective shields of a calcareous or other nature secreted from the body.

Among them we may mention, in addition to the many modifications of hair and feathers: the horny skin or shell of the tortoise; the cast scaly epidermis or slough of lizards and snakes such as may be frequently met with in the haunts of these creatures; the cast skins of frogs and newts, the delicate structure of which is easily observed from specimens preserved in a suitable fluid; portions of the skin of different fishes with large scales in situ; the shells, univalve and bivalve, of terrestrial, aquatic and marine mollusks; the calcareous tubes secreted by, and the sandtubes covering, many of the marine worms; the spiny, hardened skins of sea-urchins, etc.

As regards the homes constructed by animals, we need by no means confine our studies to the nests of birds, for we have numerous examples of beautiful and wonderfully-constructed homes without hands prepared by creatures much lower in the scale of life. Such specimens may be found in abundance, and may be collected while still inhabited, with a view of studying the habits of the occupants, or after they have been discarded by the builders concerned.

As examples we may enumerate the many different kinds of cocoons constructed by insects; the nests of various species of bees and wasps, both solitary and social; the cases very variable as regards the material of which they are constructed, of caddis-fly larvae; and the silken cocoons of spiders.

Many instructive objects, connected with both animal and vegetable life, are also to be found among the jetsam of the seashore. A single day's collecting on the shore will often provide such a store of material that many hours may be profitably spent in examining it. This material, stored in the school museum, will prove exceedingly useful at times when, owing to the weather or other circumstances, other objects are not available.

A permanent place in the school collection may be given to specimens of common rocks and minerals, including, of course, the building-stones, paving-stones, and other mineral products connected with the occupations of the neighborhood of the school.

In addition to the material mentioned above, as well as other instructive material or objects collected by the combined efforts of the teacher and the children, it will be well to exhibit, either permanently or from time to time, any good pictures which illustrate seasonal phases of Nature, together with drawings from Nature made by the children. The exhibition of the best work done by the class will prove a great stimulus to future occupations of the same kind; and, where the spirit of the work is good, such exhibitions will stimulate, rather than discourage, even those who, after many attempts, have failed to produce results up to the average of the class.

Each classroom should have a weather-chart, kept by the children under the guidance of the teacher. Such charts may be either weekly or monthly records of the weather observations made, and should always be preserved in the school museum for future reference and comparisons. Thus, it is a good plan to place old charts beside the current one, in order that the present records may be compared with those of the corresponding periods of former years, and also in order that one season may be compared with another.

Records of the observations made during the school excursions should also be preserved for reference and comparison; and any good plans or maps made by the children may be stored for the same purpose, after they have been exhibited for a period.

We referred, a short time since (page 191) to a store department of the school museum as distinct from that portion in which objects are displayed for general observation. This department consists of a number of drawers, boxes, or other suitable compartments, containing material for distribution around the class, thus providing the teacher with the means of conducting collective nature lessons. Each compartment will contain sufficient specimens, or sets of specimens, to distribute to a whole class; and since the same specimens may be used over and over again, in all the classes of the school, it will be seen that such a provision will give the teachers splendid opportunities of giving good nature lessons when outdoor observations are impossible.

The following is a list of material that may be stored for the purpose suggested:

1. Sets of dry fruits to illustrate wind-dispersion.
2. Similar sets to illustrate dispersion by animals.
3. Sets of dehiscent fruits to show their various forms and the different ways in which they split.
4. Slices sawn from an oak or other branch, with bark intact, smoothed with glass-paper on one side, for the study of the exogenous stem and its mode of growth.
5. One or more bivalve shells, such as those of the mussel, cockle, and scallop.
6. One or more univalve shells, such as those of the winkle, whelk, snail, and limpet.
7. Specimens from the seashore, such as the sea fir and sea mat.
8. Down feathers and quill feathers.
9. Specimens of common rocks: limestone, slate, sandstone, granite, etc.

10. Specimens of any minerals that are worked in the neighborhood.
11. Pieces of various metals for the study of their properties—bits of steel spring, iron sheet or wire, lead sheet and wire, copper wire, scraps of zinc plate, etc.

The above are merely a few suggestions to which the teacher may make additions according to his requirements or to the material at hand.

B. The Preservation of Specimens

We now propose to give general hints as to the simplest and best means of preparing and preserving the material required for the school collection.

Of course, a considerable quantity of this material requires no care whatever except as far as mechanical injury is concerned. Specimens of rocks and minerals and of other inorganic and indestructible matter need only be placed best side out if for general observation, or wrapped up separately in paper to prevent mechanical wear if stored away for occasional use.

This latter precaution may appear unnecessary to most readers, but one can get a much better idea of the structure of a rock specimen by the examination of a surface exposed by simply breaking it from the mass, than from a surface that has been worn down by friction or other rough usage. This is especially important in the case of crystalline rocks, in which it is necessary to note the nature of the faces of the natural crystals.

As regards vegetable specimens that are to be preserved in the dry state, the secret of success lies in a thorough and rapid drying process. Leaves and flowers should be arranged among sheets of blotting or other thick porous paper, always allowing several sheets of paper between each two adjacent layers of leaves, etc., and then placing the whole between two boards with a heavy weight on the top. If the specimens contain plenty of sap it will be advisable to turn them out after several hours, and then replace them, as before, using a fresh supply of paper. In fact, it is a good plan to keep the press in two parts, and use them alternately, the paper of the one part being spread out to dry while that of the other is in use. If the drying process is slow, more or less decomposition will set in, destroying the natural colors. If, on the other hand, the drying is rapid and thorough there is much more chance of preserving the original tints.

There need be no great expense attached to the making of a good botanical press, for almost any cheap, thick, unglazed paper will serve for the purpose. Perhaps one of the best and cheapest is the coarse sugar paper used by the grocer.

When it is required to store the dry fruits of forest trees, plants, etc., simply spread them in the sun or before a fire until they are quite dry. If stored while there is still moisture within them they will soon be attacked by mildew or other parasitic growths, and their natural appearance and their usefulness quite destroyed.

After vegetable specimens have been thoroughly dried, they may still be attacked by mildew if stored in a damp place, or by insects and other museum pests if not secured in air-tight boxes or cases. To lessen the risk of such destruction, sprinkle them with a solution of corrosive sublimate in methylated spirit, and store them as soon as the spirit has evaporated. It must be remembered, however, that this solution is a very poisonous one, and should therefore be kept out of the reach of children.

It will seldom be found necessary to preserve soft and succulent vegetable specimens. These are always more suitable for study while they are fresh and in season. Should it be desired, however, to preserve any special kinds for future reference, they may be put in jars or bottles of methylated spirit or dilute formalin.

The latter preservative is, perhaps, the better for general use. It is usually to be purchased as a 40%, solution, but a 2% solution is quite strong enough for all purposes, and this latter may be made by taking one ounce of the original liquid as purchased, and increasing the volume to one pint by the addition of water.

Formalin is also the best general preservative for all kinds of animal structures, and it will maintain them in such a condition, for an indefinite period, that they may be removed from the liquid at any time for detailed examination. It is suitable for even the softest and most delicate objects, such as jellyfishes, and the cast skins of amphibians.

Animals of a harder nature, such as sea-urchins, starfishes, small crustaceans, and insects with hard coverings, though perfectly preserved by spirit or formalin, are frequently best stored in a dry condition, properly mounted and secured in glass-topped boxes. They should always be dried rapidly, in a warm, airy place; and the drying process is much more rapid, and, as a rule, much more satisfactory in every respect, if the specimens have been previously soaked in methylated spirit for several hours. The spirit extracts the moisture from the softer internal structures, causing them to contract and dry rapidly; and if a little corrosive sublimate (mercuric chloride) be dissolved in the spirit employed for the purpose, the dried specimens will never be attacked by any of the troublesome museum pests.

Many museum specimens, such as feathers and other coverings of animals, are liable to be attacked by the larvae of small moths. If such specimens are to be mounted for display, and not secured in well-fitting glass-topped boxes, they should be wetted with a weak solution of corrosive sublimate, and then thoroughly dried before mounting. If, on the other hand, they are merely to be stored until required for distribution to the children for examination, they may be kept in boxes or drawers together with a few pieces of albo-carbon. This substance, which may be obtained at a low price from any ironmonger, will effectually prevent moths from intruding and depositing their eggs; but, since it is a volatile solid, the supply should be renewed at intervals of a few months, or as found to be necessary.

We have previously referred to the value of the skulls and bones of animals for teaching the nature of the different kinds of teeth and the structure of the different types of joints. Such useful specimens are often to be found among the refuse of the kitchen. They are easily cleansed by means of hot water containing a little washing soda, and may be bleached to whiteness by a prolonged exposure to the summer sun.

Many very valuable specimens of this nature are to be found during country rambles. The skulls and, in fact, the whole skeletons of birds, rabbits, and other animals, are often met with in the country—skeletons that have been cleaned perfectly by carrion-feeding insects and beautifully bleached by exposure to the sun's rays. These will prove very instructive in the school.

Whatever be the amount of care bestowed on the preparation of animal and vegetable specimens for the school museum, it must be remembered that the objects are always liable to the attacks of moth and various minute marauders. Hence it is necessary, at intervals, to thoroughly examine every mounted specimen and the contents of every store-box or other compartment, giving special attention to the material that is not frequently used.

If any marauders are present, transfer the specimens attacked to a box with a perfectly-fitting lid, sprinkle them with benzole or chloroform, and shut them up securely for some hours. The pests having been thus killed, the specimens may be moistened with a weak solution of corrosive sublimate in spirit, dried, and then returned to their proper places.

C. The Arrangement and Labeling of Specimens

Little need be said concerning the arrangement of the objects displayed in the school museum. It is advisable to utilize the whole of the glass-fronted space available for the exhibition of the specimens that are of such general interest as to attract young observers during their spare moments, and not allowing this space to become overcrowded by using it as a store for material that is required only occasionally for class teaching, and which is more satisfactorily housed in ordinary store-boxes or drawers.

Every displayed specimen should be at a convenient height for observation, well-lighted, and visible without any obstruction.

The arrangement need not necessarily be made with regard to scientific classification. In fact for school purposes it will be far better to arrange the objects on a seasonal basis, if there is to be any arrangement at all, except as regards the sizes of the objects and the spaces available.

Many teachers may have noticed that a school museum in which objects are always displayed for observation, does not attract the attention of the children as it was hoped to do, especially if the same specimens are always on view. The museum soon becomes a familiar object, looked upon merely as an item of the school furniture, and is consequently ignored.

In order to prevent this some arrange to have the specimens screened as a rule, and displayed to view only on certain occasions, during which a few remarks may be made by the teacher with the intent to arouse the interest of the children. A still better plan, where the collection of material is sufficiently extensive for the purpose, is to change the specimens on view at fairly regular intervals, giving each time some general account of the exhibits to stimulate careful observation.

On the other hand, we are of opinion that in those schools where nature study is regularly and properly conducted, and where the children themselves are the prime movers in the formation of the school collection of natural objects, there will be no need to resort to stratagems for the purpose of arousing the desired enthusiasm on the part of the children; and the very frequent additions to our museum will always be sufficient to maintain a small crowd of anxious, inquiring, and admiring spectators. The main difficulty will be, it seems to us, to find sufficient accommodation for all the interesting things that have been brought in for observation.

The matter of labeling is one of great importance. A museum has been described as an interesting collection of labels illustrated by specimens; and, whether this definition be accepted or not, we must insist that every specimen is at least accompanied by a label stating its name, date of collecting, locality, and, to give due credit and encouragement to the young naturalist concerned, the name of the donor.

EGG OF DOG-FISH.
FOUND ON THE BEACH BY
WILLIE SMITH
AUGUST 1910.
What is the use of the curly tendrils?

A specimen label for a school museum (Fig. 182)

But this alone is not sufficient for a school museum. We want to encourage the closest possible observation and the spirit of thoughtful inquiry. In order to do this we may place on the label some instructions, stated, of course, in a general way only, as to what is to be seen. The children should be told, without any particularizing, what they should look at, and what they should look for; and a question, also written on the label, will often lead to an inquiring state of mind that must necessarily work for good. We ourselves have seen a group of youngsters, stimulated by a question written on a museum label, engaging in a very interesting discussion on the problem raised—a discussion rendered all the more valuable on account of the great diversity of opinion expressed. This is just the kind of thing we want to encourage.

Finally, we desire to emphasize our few remarks on the labeling of specimens, not only because, on the one hand, we are convinced of the usefulness of labels in encouraging good work, but, on the other, because, though we have seen a large number of school museums, we have never yet met with a single one in which a good method of labeling had been systematically carried out.

— 13 —

The School Aquaria

The value of school aquaria can hardly be over-estimated, for the creatures of our ponds and streams exhibit a wonderful variety of structure and habit, and equally wonderful adaptations of structure to habit and resemblances to the environment. Many of them, too, undergo a remarkable metamorphosis that is very easily observed.

We propose, in the present chapter, to give some general hints as to the establishment and management of aquaria, and to dispose of the imaginary difficulties that are so often supposed to exist in connection with this department of the nature work.

First then, as regards the vessel or vessels to be employed, it should be stated at once that expensive aquaria of the patterns most frequently sold are not at all necessary.

If a rectangular tank is desired, a glass accumulator cell or tank is by far the cheapest kind. True, the slight unevenness of the glass interferes slightly with the vision of objects viewed through the sides, but many of them are so well made that this objection is hardly appreciable. The tanks we refer to are to be obtained from almost any manufacturing electrician, and may often be secured, at a very low price, from dealers in second-hand wares.

The well-known bell-jar supported on a stand, is also useful, as well as the fish-globes that require no stand, but they have the objection that the glass is frequently very thin and easily broken.

Very large aquaria are required for fishes only. For aquatic insects and other small aquatic creatures vessels of almost any size will suffice, down to a capacity of a quart or less.

Even if a very large aquarium has been selected, it must be remembered that the one is not sufficient. Not only would it be difficult to keep small things under proper observation in a large vessel, but the accommodation must be such that the carnivorous creatures are always separated from the harmless things they would devour. The smaller animals, such as bloodworms, the larvae of gnats, etc., may be kept in any kind of wide-mouthed bottle, glass flask, or even tumbler.

It is often stated that the top of the aquarium should be very wide, in order that a large surface of water may be exposed to the air for the absorption of oxygen. This advice is good, from a general point of view; but it is possible, as we shall see, to stock an aquarium in such a manner that we are practically independent of the atmosphere, and therefore do not need such a large water-surface.

Again, we believe that failure to keep a school aquarium in good order often arises from the fact that the water-surface is large, and therefore capable of collecting such large quantities of dust, especially while the schoolroom is being swept. If an aquarium is well supplied with growing aquatic vegetation, sufficient to produce the oxygen required by the animal inmates, a small mouth is an advantage rather than otherwise for a school aquarium, especially as concerns those schools in which the sweeping operations are so conducted as to fill the air with poisonous dust.

Our advice, then, is briefly this: Secure at least one large vessel—the larger the better, for fishes; and several smaller ones for the accommodation of the small forms of pond life. Arrange them all in a good light, but not where they will receive the full force of the summer sun, and all at a convenient height for observation. If there is not window space for them inside the school, place them on shelves outside, preferably against a wall facing the north, so that they shall not become heated by the direct rays of the sun. Of course these outside aquaria must be brought inside during frosty weather, or the formation of ice may lead to damage.

Permanent outdoor aquaria are very useful, especially for the purpose of keeping fishes, and other rather large aquatic animals such as crayfishes and pond-mussels, also for the growing of large aquatic plants. A large tub will make a very satisfactory outdoor aquarium. A very useful one may also be made by constructing a large, strong, wooden box, say about four feet long and wide, and fifteen inches deep, and fining this with a thick layer of Portland cement mixed with about an equal quantity of sand. Such an aquarium might be made by the children themselves. The manual work, under proper guidance, will be a good training for them, and they will also be able to learn something concerning the properties of cement. If the cement cracks during the drying, as it probably will, the cracks may be filled by running into them some neat cement made very thin.

Having procured the utensils necessary, the next thing is to introduce some kind of soil for the purpose of holding the roots of the aquatic plants. Large outdoor aquaria should have a layer of clayey soil, a few inches deep. Indoor aquaria may have a corresponding quantity of well-washed coarse sand; but if they are to contain aquatic larvae or other creatures that find their food in mud, then a little mud from a neighboring pond may be substituted for the sand.

Next, introduce some aquatic plants. These, under the influence of light, absorb the carbonic acid gas produced by the respiration of aquatic animals, and, after decomposing this gas (which is an important plant food), set free oxygen. Thus they tend to keep the water in a proper condition for animal respiration. The pondweeds also add considerably to the general appearance of the aquaria, and, in some instances, provide food for the animals.

Some aquatic plants are far more suitable to our purpose than others, and preference should be given to those species which are easily propagated, grow rapidly, are of suitable size, and give off liberal supplies of oxygen. Those especially recommended are the water starwort (*Callitriche*), hornwort (*Ceratophyllum*), American pondweed (*Anacharis* or *Elodea*) and the river grass (*Vallisneria*), while the little floating duckweeds are also pretty and useful.

Not only is plant life essential to the well-being of the aquarium, as we have already shown, but aquatic plants are in themselves very useful objects for study, their growth and habits being peculiarly interesting.

They may be planted in the aquarium after the latter has been filled with water, their roots being simply pushed into the sand or soil by means of a stick. Some species, such as the starwort and the American pondweed need not be fixed at all. They grow well without any soil, and where soil or sand exists the plants will develop roots and fix themselves. Indeed, if mere fragments of the plants we have named are simply thrown into the aquarium they will thrive almost as well as if they were complete plants properly rooted.

Now, with regard to animal life for the aquaria, it is most important that there be no overcrowding. The water-breathing creatures introduced should be so few in number that they require no more oxygen than is liberated by the vegetation present. If more than this, it will be necessary, at more or less frequent intervals, to change the water, but it is far more satisfactory to keep such a balance of animal and vegetable life in each vessel that a change of water, with its accompanying disturbance of the living things, is never necessary. A well-balanced aquarium ought to remain in a perfectly satisfactory condition for several years without a change, the only attention, as far as water is concerned, being an occasional addition to make good the loss by evaporation.

No more than one or two small fishes should be put in an aquarium for each gallon of water contained; while various small animals that obtain their oxygen from the water, such as water snails, leeches, and some aquatic larvae, may be introduced more liberally, since they require but little oxygen for respiratory purposes.

When estimating the amount of animal life that may be accommodated in each aquarium, it is well to distinguish between those creatures which take dissolved oxygen from the water, and those which, while they live wholly in water, come to the surface for all the air they require, and therefore do not tend to render the water unfit for the respiration of the water-breathers. This difference in the manner of breathing is easily distinguished by careful observation.

It will be observed, for example, that water beetles, water bugs and water spiders always come to the surface for the necessary air, and carry their supplies down with them, while some of the aquatic larvae thrust their breathing tubes or appendages above the surface, or even suspend themselves from the surface-film of the water by means of their breathing apparatus. Water snails and some aquatic larvae, on the other hand, always remain below the surface, and obtain all their oxygen from that dissolved in the water.

The fish usually selected for aquaria is the well-known goldfish, but our ponds and streams will provide even more interesting, though not so gaudy, species, concerning which a few remarks have previously been made (page 75).

When collecting pond life for the aquaria—a work that should be conducted by the aid of a shallow gauze or muslin net with a strong wire frame and long handle—make it a rule to reject no species found. There is no need to reject an animal because it is not known. It will provide as useful study as any other, and there is great pleasure in making oneself acquainted with new things.

Following this plan, however, a productive pond will possibly yield more than can be properly attended to; but make a selection, and sort out the different forms of life, placing the different species in separate small aquaria for observation. A little observation and experiment will soon enable one to determine what food is necessary, and it will often happen that a creature, apparently uninteresting in habit, will turn out to be particularly instructive, and, perhaps, to undergo remarkable metamorphosis.

If an aquarium is placed in a strong light, such as in a window facing south, it will often happen that the glass becomes covered with a green deposit of a low form of vegetable life, sometimes so dense that the animals within are observed only with difficulty. When such is the case, there is no need to change the water or to clean the glass. This vegetation, though it obscures the observation, tends to keep the water in perfect condition for the animal life. Whenever it is necessary to reduce it, simply place a screen of more or less opaque paper between the aquarium and the window, and the green deposit will gradually disappear. Sometimes the whole of the water will become charged with floating, green vegetable cells, but they may be reduced in the same way.

Water snails are also useful for helping to keep the glass sides of the aquarium free from vegetable deposits. Some species will spend practically all their time in creeping over the glass, feeding on these deposits as they go, rasping the same from the surface by means of their hundreds of minute teeth. Some creatures, too, are extremely useful as aquarium scavengers, since they feed on any decomposing matter that settles on the bottom. Perhaps the most useful of these is the little crustacean known as the water hog-louse (Fig. 137).

For the purpose of studying the metamorphosis of aquatic insects we strongly recommend the bloodworm, which develops into a gnat-like insect, the larva of the gnat itself, and the larvae of may flies, caddis flies and dragon flies. The last of these is carnivorous, while the others feed on material which they find in the mud of the pond or stream in which they live. They may be placed in aquaria about half filled with water, with a few sticks or some other means by which they can climb out of the water when about to undergo their final change; and muslin may be tied over the top of the vessel to prevent the escape of the perfect insects when they appear.

The water snails will often deposit their spawn—eggs enclosed in a gelatinous mass—on the glass, and thus give splendid opportunities for watching the development of the young by the aid of a magnifying lens.

We have now to refer to a matter of very great importance to the aquarium-keeper—a matter in which, we believe, the most serious of all mistakes is made. It concerns the feeding of the captive animals. We believe that more failures are caused through overfeeding than by any other error, not that the animals devour more than is good for them, but because the excess of food decomposes, rendering the water putrid and poisonous.

Make it a rule, then, to give no more food than is found to be absolutely necessary; and if the food given is not all devoured in a very short time, remove what is left without delay. Any small fragments at the bottom may be removed by means of a glass tube, used after the manner of a pipette. If the water ever becomes turbid, and emits even a slight odor of putrefaction, run it all off, and fill up the aquarium afresh; but this will never be necessary in a well-managed aquarium, kept according to the advice given above.

Salt-water aquaria should be kept in schools situated near the sea, for the closer observation of the common animals seen on the shore; and there is no reason why such aquaria should not be established also in the schools of inland towns. They are not so easily managed as fresh-water aquaria, mainly on account of the fact that the marine plants do not give off oxygen in such abundance as many of the fresh-water species, thus making the proper aeration of the water a problem of greater difficulty.

Great care must be taken not to overcrowd a marine aquarium with animal life; and there will generally be a necessity for rather frequent changes of water unless some mechanical means of aerating it can be devised.

Fresh sea water can now be obtained daily in nearly all inland towns; but, failing this, a substitute may be prepared in the form of an artificial sea water, made by dissolving sea salt (not table salt) in fresh water in the proportion of seven ounces of the former to ten pints of the latter.

Before leaving the subject of the management of aquaria we feel it necessary to make a few special remarks concerning the rearing of frogs and other amphibians from the spawn. And we do this not only because we regard the metamorphosis of amphibians as peculiarly interesting and instructive, but also because of the many difficulties and failures that have been brought to our notice, and of the numerous questions that have arisen in consequence.

Frog eggs are to be found in almost every pond and ditch during the latter part of March and in early April. The mass of spawn laid by a single frog is very large after the gelatinous coverings of the embryos have absorbed water, and it consists of hundreds of eggs all joined together.

It is a very common mistake to put this whole mass of spawn into an aquarium of a medium or small size, with the result that the air supply for the hatched tadpoles is quite insufficient; and further trouble arises when the vacated gelatinous matter decomposes, rendering the small amount of water quite putrid and, perhaps, causing the whole colony to perish.

If it is desired to rear the complete offspring of the frog, a large aquarium must be used—one holding several gallons of water, and it must be very liberally supplied with aquatic plants; but if such an aquarium is not available, divide the mass of spawn by cutting through it with a blunt knife as it lies in a shallow dish of water, and never put more than a score or so of eggs for each gallon of water the aquarium holds.

At first the tadpoles feed entirely on vegetable matter, and they may be seen rasping away the tender leaves of the plants with their horny jaws.

The aquarium should be in a good light, and it will be all the better for a few hours (not more) of direct sunlight during each day. The strong light will encourage the development of a green vegetable deposit on the glass of the aquarium, and the young tadpoles will rasp this away with their jaws, thus giving a splendid opportunity of observing the manner in which they feed.

After a few weeks they require animal food, and will then readily attack almost any kind of dead animal matter, such as a dead worm or a piece of meat. It is important to observe, however, that no more animal food must be supplied that they actually require, for any excess, left in the aquarium, will become putrid, poison the water, and cause the death of all the inmates.

A small piece of tender meat may be suspended in the water by means of a thread—a plan that allows the food to be changed and renewed without any difficulty.

As the lungs of the tadpole develop, the latter may be seen coming to the surface for air, and it is then necessary to float a small sheet of cork or a piece of thin, light wood, in order that the creatures may rest at the surface as they require.

It is important to note, however, that as soon as the young frogs show a decided tendency to leave the water, they should be removed from the aquarium, and either placed in a vivarium, the bottom of which is covered with damp turf and which contains a shallow vessel of water, or else set free in the school garden or on some neighboring grassland.

We do not recommend the keeping of very young frogs in a state of captivity unless the keeper is prepared to spend a great deal of time in providing them with the proper food. At this period they require a large number of very small insects, together with very minute worms and slugs that they are capable of swallowing whole; and the collection of this food will often prove so difficult, or will take up so much time, that there will be some danger of the loss of the captives by starvation.

If, on the other hand, the young frogs be set free on a patch of grassland, where there is some kind of cover in which they can hide when the day is bright, they may still be seen occasionally, and their progress noted.

The life history of the toad is practically the same as that of the frog, and the same treatment is necessary in rearing the tadpoles. The eggs of the toad are usually deposited a little later than those of the frog. They may be seen in ponds during April, not in rounded masses like those of the frog, but in very long strings, entangled among, and supported by, the pondweeds.

Newts commence to deposit their eggs in early spring, and often continue the operation of egg-laying for weeks or months. The eggs are laid singly, usually on the underside of the leaf of an aquatic plant; and the female often curls the leaf on which an egg is deposited, by means of her hind legs, apparently to protect the egg from being devoured.

If a few pairs of newts be placed in a large aquarium, furnished with an abundance of pondweed, during the early spring, the process of egg-laying may be observed, as well as the metamorphosis of the young.

While the tadpoles of the frog and the toad are very similar in their appearance and metamorphosis, those of the newts are different in many respects. The latter are of a much lighter color, with bodies more elongated. Their external gills persist much longer, the fore legs appear before the hind ones, and the tail does not disappear.

— 14 —

The School Vivaria

In this chapter we shall include the appliances necessary for the keeping of reptiles, amphibians, mollusks and a few other kinds of animals, leaving the directions for the management of mammals and birds, and for the rearing of insects, to be dealt with later.

A reptile case should be large, with a front and back of glass. A convenient size is 3 ft. long, 1 ft. wide, and 18 in. high. A good one may be made, at a very low cost, with bottom and ends of wood, grooved to receive the glass back and front, and a top of perforated zinc that is quite secure as far as the escape of the reptiles is concerned, but easily lifted when occasion requires.

It should be placed in a window where it can receive direct sunshine. The bottom may be covered with coarse, clean sand, or shingle, strewn with some loose moss; and a few branches or twigs should be introduced in order that the creatures may be able to climb. A shallow vessel of water is also necessary for the animals to bathe in and drink.

Such a case may contain either or all of the British lizards—the common lizard (*Lacerta vivipara*) which may be caught on heaths and banks in most parts of the country, the sand lizard (*L. agilis*) which is frequent on sandy heaths, and the blind-worm or slow-worm (*Anguis fragilis)*, a snake-like lizard without legs, frequent on heaths and commons and along the borders of woods. A few foreign species are also very suitable for observation in captivity, and among these we may particularly mention the green lizard (*Lacerta viridis*) and the wall lizard (*L. muralis*), both of which are well-known continental species that are frequently seen on sale in the naturalists' shops of our country.

The common English ringed snake or grass snake also makes an interesting pet; but, as we have previously stated, the venomous viper or adder should not be introduced into a school vivarium unless it is thoroughly understood by the teacher in charge; and, if kept, should always be secured in a locked case of which the teacher has charge of the key.

Both snakes and lizards may be housed in the same case, but it is better to keep them separately, since the grass snake will sometimes swallow a lizard.

The lizards feed on small worms, slugs, caterpillars and various other insects, and spiders; while the favorite food of the grasssnake consists of frogs and newts. In both cases the prey must be presented alive, for neither lizards nor snakes will, as a rule, touch anything that is dead or motionless. The water should be renewed frequently—at least two or three times a week.

It must be remembered that both lizards and snakes are good climbers, and thus it will be necessary to see that there are no crevices or other spaces in any part of the vivarium through which the creatures can escape, bearing in mind, too, that they can squeeze themselves through openings which appear much too small to admit their bodies. It is a convenience to have a hole, guarded by a sliding door, through which the food may be introduced, so that it is not necessary to raise the top every time the creatures are to be fed.

Under their natural conditions, snakes and lizards hibernate throughout the cold season—from October to March—taking no food, and if, in captivity, they are placed in a cold room during this period, they will remain in the same state. If, however, the creatures are kept in a warm room during the winter months, they will remain more or less active, and will require an occasional meal.

The food given should never be larger than the reptiles can swallow whole, since the latter are unable to divide it with their small teeth. Furthermore, the food should not be provided in excess, for the undevoured prey will, sooner or later, die in the vivarium, and the decomposing bodies will give rise to unpleasant and insanitary conditions. Feed the animals as long as they are eager for food, and remove all excess of food.

At times it will be necessary to clean out the vivarium. For this purpose, remove all the reptiles and turn out the shingle and moss. Then clean the glass, and supply fresh shingle and moss, or return the same after it has been well washed and dried.

The reptiles will soon become very tame if properly cared for, especially the lizards, which will, before long, accept their food from the hand. In this condition, too, they may be handled freely for closer observation.

A vivarium constructed of wood and glass, similar to that suggested for reptiles, will also suffice for the amphibians—frogs, toads and newts—but one made entirely of metal and glass is better for the latter creatures, since it must always be kept damp. If wood is used for the bottom and ends it should be painted (two or three coats) inside as well as outside. A shallow dish of water should be provided; and the whole of the bottom, except the space occupied by the dish, should be covered with a large turf. A little shelter, made by loosely piling up some stones in one of the corners, will also be a useful addition.

The turf should be occasionally sprinkled with water if it becomes dry, and the water in the dish will require to be changed at intervals of a day or two. A fresh turf will also be supplied as appears necessary.

The amphibians will require small worms, slugs, and various insects as food; and frogs, toads and newts may all be kept in the same home.

It should be observed that the adult animals of these species prefer a pond as their home during early spring—the breeding season. At this time it is better to keep the frogs and toads in a garden pond (page 228), and the newts in any kind of aquarium. Later in the year, when the deposition of eggs is over, they may all be transferred to the vivarium as above described.

Snails, slugs, centipedes, millipedes, beetles, and various larvae that are to be found in the soil may be kept for observation in almost any kind of case with one or more glass sides and a movable top of perforated zinc. Any wooden box of suitable dimensions may be adapted to this purpose by replacing one side with glass, and fastening perforated zinc over a large hole cut in the lid.

Speaking generally, there is little need of keeping such creatures as those above mentioned in captivity, for they can be so easily observed outdoors. There are times, however, when it is advisable to retain them for a period for close or continuous observation in order that their growth may be watched, their habits more accurately determined, the development of the eggs observed, etc.

A vivarium intended for the above purpose should have a rather deep layer of soil at the bottom, and a shelter of some kind provided for the creatures which require it. The soil should be kept damp by an occasional sprinkling of water, and renewed when necessary. The proper food should, of course, be regularly supplied, and any excess removed in order to prevent the unpleasant and dangerous results that may arise from putrefaction.

Many of the creatures that inhabit the soil of the garden are nocturnal, and, consequently, will remain under cover throughout the day. It will be necessary to turn them out of their hiding places if they are to be observed during the daytime, and to watch them with the aid of artificial light after dark. Many also are of carnivorous habit, like centipedes, some millipedes, and numerous beetles and their larvae, preying on the weaker creatures which they capture, while some feed on decomposing organic matter, as the carrion beetles and the grubs of various flies.

It will thus be seen that many of our garden creatures have habits that render them unfit for an indoor life of captivity, yet there are times when it is convenient to house them for a short period to carry out certain observations that would present difficulties outdoors.

In order to study the interesting habits of earthworms it is necessary to prepare a special kind of vivarium. Procure or construct a wooden box about 1 ft. long, 6 in. wide, and 1 ft. high, but with one of its broad sides (1 ft. square), which is to be the front, of glass for observation. Perforate the bottom with several very small holes for drainage, and make a well-fitting lid of perforated zinc for the top. None of the perforations, either in the top or bottom, should be sufficiently large to allow a worm of moderate size to pass through, nor should there be any cracks or crevices large enough to allow its escape.

Fill the box, to within about two or three inches of the top, with two or three different kinds of soil, arranged regularly. Thus, at the bottom there may be a few inches of a dark-colored loam, above this about as much yellow loam, and above this again a layer of dark, peaty soil. Each soil should be sifted before it is introduced, and pressed down rather compactly with a block of wood so that its surface is level and even, and the soil itself free from spaces of any appreciable size.

When the soil is ready, scatter a few small fragments of dead leaves on the surface of the upper layer, introduce several large earthworms, and sprinkle the soil with water unless already rather damp. In any case a sprinkling of water will be necessary at frequent intervals, for the soil must always be in a damp condition.

In a short time some of the burrows of the worms will be seen close to the glass, worm castings will be observed on the surface, and fragments of dead leaves that have been pulled down into the burrows.

— 15 —

The Rearing of Insects

Few natural history studies are as fascinating as that of the metamorphosis of insects; and in proportion as the changes which these creatures undergo excite wonder and surprise, the study of them is valuable as a means of creating a desire to search into the marvelous ways of Nature in all her aspects.

Moreover, this branch of the nature work is particularly easy of accomplishment, for the material is always to be found readily either in town or in country, and there are but few difficulties that stand in the way of success.

The rearing of any one species may be started with the eggs or with the larva or grub; and preference will be given, as a rule for obvious reasons, to herbivorous rather than to carnivorous species, especially if the rearing is to be conducted indoors.

If we start with the eggs it will be well to remember that the female insect generally deposits these on the food plant or food material that the larvae require; also that while some grubs will eat only one particular kind of food, others are less restricted in their habits in this respect. For example, while the larvae of the pretty tortoiseshell butterfly (*Vanessa Urticae*) feeds only on stinging nettles, those of the well-known painted-lady butterfly (*V. cardui*) may be found feeding on mallow, burdock, cud-weed, viper's bugloss, and several species of thistles.

If a cluster of the eggs of an insect be discovered on the twig of a tree or on the branch of a plant in the garden, and it is desired to find out what they are and how they develop, enclose the twig or branch in a muslin bag, the mouth of the latter being tied securely around the base of the former. When the young larvae emerge from the egg they will have sufficient food within to last them for some time, and the bag may be carefully removed at times in order that their progress may be better observed; also, when they have devoured all the food enclosed in the bag, they may be carefully transferred to another branch and secured as before.

Of course it is not absolutely necessary to imprison the grubs at all, but the probability is that if they have full liberty there will often be some difficulty in finding them. It is almost certain, too, that many of them will be devoured by insectivorous birds if left exposed, or attacked by ichneumon flies and other parasites.

Larvae secured on the growing plant or tree as above described may be so treated until they are fully grown and ready to change to the chrysalis or pupal condition, but while some species always descend and burrow into the ground when about to undergo this change, others protect themselves by silken or other cocoons, or secure themselves by means of silken cords or webs, and change above ground. If the habits of the species under observation are unknown, or if it is desired to keep them indoors for closer examination, the larvae may be transferred to a suitable cage such as will be presently described.

Sometimes the eggs or caterpillars of a certain butterfly or moth may be seen on the leaves or stem of a low-growing plant in the garden, in which case, supposing it is desired to rear the insects outdoors, the whole plant may be enclosed in a muslin bag tied around the base of the stem. If it happens to be a species which is known to burrow into the ground when about to change to the chrysalis, the muslin bag, instead of being tied around the stem of the plant, may be sewn to a ring of wire, and pinned down close to the ground by means of pieces of wire bent into the form of hair-pins.

Then, after the change to the chrysalis has taken place, the insects may be dug out of the ground and transferred to a box kept in the schoolroom, in order that the perfect insects may be observed, and, if circumstances are favorable, even the act of emergence from the chrysalis case. The caterpillars of the large white butterfly (Fig. 41), the small white butterfly and the cabbage moth (Fig. 208), may all be reared outdoors in the manner described. The first two of these will probably fasten themselves to the muslin bag by means of silken cords when they are about to change to the chrysalis, and when they have all changed they may be brought indoors, still enclosed in the bag, in order that the final metamorphosis may be observed. The last-named insect burrows into the ground before changing.

All three of these insects are commonly seen on cabbages where they prove themselves very destructive, but they also feed on other plants of the same order.

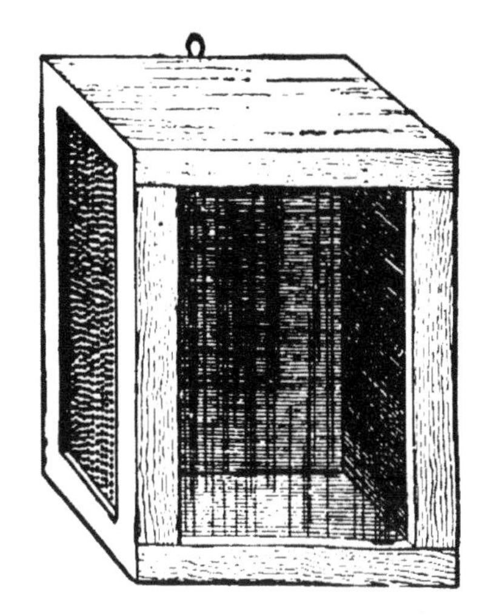

A cage and a bell-jar for raising insects (Fig. 183 & Fig. 184)

Cages for rearing insects indoors may consist of boxes with glass fronts, movable lids or doors, and finely-perforated zinc at the top, back, or sides to allow of ventilation. About an inch of dry, sifted soil should be placed in each box, and the food plant which, of course, should always be renewed as it is devoured or loses its freshness, may be fixed in a small jar or wide bottle filled with wet sand. The food plant should not stand in a vessel of water unless there is some arrangement for preventing the insects from falling into the water, or several may be lost by drowning.

A very healthy cage for the rearing of caterpillars may be made by fixing a cylindrical muslin bag, with a draw-string at either end, to two rings of wire. Such a cage may be suspended by one draw-string, while the other is tied around the neck of a bottle of wet sand that supports the food plant. The upper end, of course, serves the purpose of a door for the introduction of insects or food (see Fig. 185).

Whatever be the form of cage used, cleanliness must be observed. All waste food and all excrement must be removed at suitable intervals, and, if there is a layer of soil at the bottom, this also should be renewed occasionally.

We have already referred to the rearing of aquatic insects such as gnats, caddis flies, dragon flies, etc., in our chapter on the school aquarium.

The observation of ants and their wonderful habits is extremely interesting. The large wood ants, the hills of which are so frequently seen in woods and on shady banks, are best observed outdoors, but the smaller garden ants may be very successfully kept and reared in an artificial glass nest in such a manner that all their movements, habits and metamorphosis may be studied in the schoolroom.

There are two very common species of these smaller ants, the black and the yellow, both to be found in our gardens and, in fact, almost everywhere. The latter is, perhaps, the better for keeping in confinement.

An artificial ant nest may be made as follows: Cut a slab of wood about 10 in. square and half an inch thick. Fix all around the edge of this, on the upper surface, by means of small brads, a border of wood having a cross-section of a quarter of an inch square.

Cut a piece of window glass that will exactly fit into this raised border, and drop it into its place. Cut strips of felt or cloth about a twelfth of an inch thick, half an inch wide, and place these all around the glass, on its upper surface, forming a continuous border; then cut a second piece of glass, the same size as the first, and lay it on the felt. The nest now consists of two plates of glass, separated by means of the felt to such a distance that there is only just room for ants to move about freely between them, and the whole contained in a very shallow wooden box.

Next, cut away about a quarter of an inch of the wood border, preferably at the middle of one end of the frame, and also divide the felt at the same place, thus leaving a small doorway for entrance and exit.

A Muslin Cage for raising insects (Fig. 185)

Finally, fill the space between the two sheets of glass with fine sifted soil that is very slightly damp.

The nest is now complete, but it should be enclosed in a shallow box, about two or three inches longer than the nest itself, and only about an inch deep. The box must be covered with a glass plate, but the latter must lie so truly on the former that there is no possible escape for the ants. The best way to make it quite secure is to glue strips of velvet all round the edge of the box for the glass to rest on. All is now ready for the introduction of the ants, which may be carried out thus.

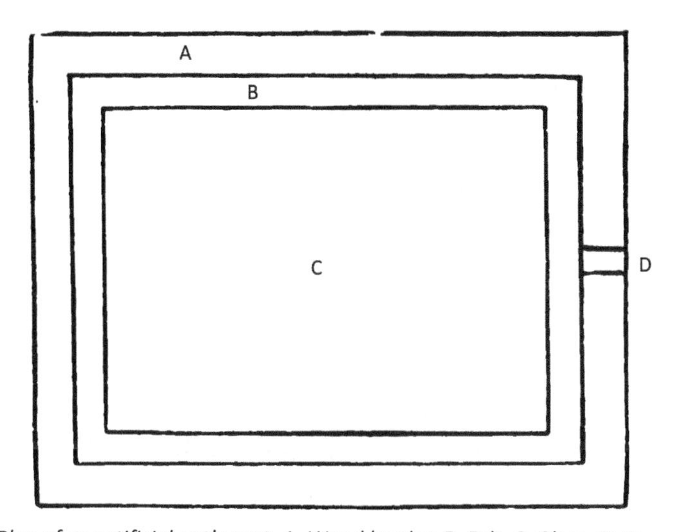

Plan of an artificial ant's nest. A. Wood border, B. Felt, C. Glass, D. Door

Section of the same ant's nest (Fig. 187)

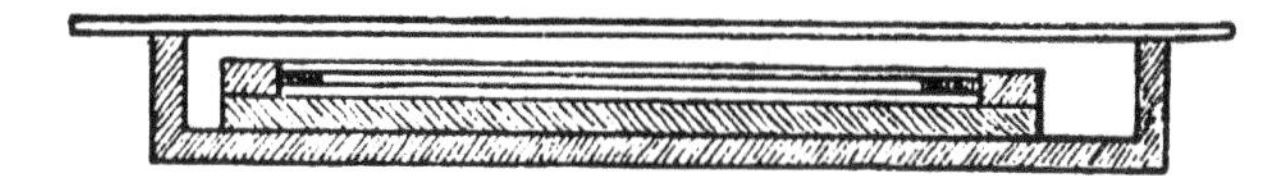

The nest enclosed in a shallow box with a glass lid (Fig. 188)

Remove the covering glass of the box, and lay a sheet of brown paper on the upper glass of the nest, the paper being of about the same size as the glass. Now search for an ant nest in the garden or—anywhere, being provided with a small trowel and a tin box with lid. The whereabouts of a nest may be soon ascertained by watching the movements of ants, and it will probably be found under the cover of a large stone. Expose the nest suddenly, and scoop up, by means of the trowel, as many ants as possible, together with their larvae and eggs (pupae), but not much soil. Throw the whole into the tin box, and close the latter quickly. Now turn out the contents on the sheet of brown paper that covers the artificial nest, and quickly replace the glass cover of the box.

The ants may now be watched through the glass cover. They will be seen hurrying in all directions, carrying eggs, larvae and pupae, and searching for a suitable home in which to settle down and restore order.

At first they will probably endeavor to establish a new home among the earth that was thrown in with them, and they may be satisfied with this arrangement for many hours, or even days; but sooner or later the soil will become too dry, and the ants will not be content with the imperfect protection from light afforded by the small amount of soil. So, at last, they will convey all their eggs and young to the prepared nest between the two glasses where the darkness and dampness are more congenial.

When this is done, remove the brown paper and the soil on it, replace the upper glass, and cover the whole with a sheet of brown paper to keep the interior dark. The paper may be removed at any time for observation, but should always be replaced as soon as the observation is over. The upper glass may also be slid along slightly in order to introduce food, and, occasionally, a sprinkling of water to maintain a slight dampness within.

The ants may be fed on fruit, small pieces of meat, crumbs of bread, small insects, honey, etc., these materials being dropped into the box at the space left for the purpose, the doorway of the nest being, of course, at this end.

With the arrangement above described it will be possible to observe all the movements and metamorphosis of the ants; and the same colony of insects, if properly treated, may be maintained for a very long time.

The observation of hive-bees is, of course, highly instructive; and while the study of bees is especially valuable in country schools, since beekeeping is so often a very profitable occupation of villagers, there is no reason why hives should not be established in the gardens or playgrounds of town schools, where they are often very successful.

If the teacher himself is not experienced in the management of bees, he will generally, at least in the country, be able to secure the assistance of a neighboring beekeeper in allowing himself and his class to witness the interesting manipulations that form the principal part of the beekeeper's work.

Observation hives may be purchased (or easily made) that will allow a view of the movements and development of the bees within; or any ordinary box-hive may be converted into an observation hive by cutting a large piece out of one or more sides, fitting the holes with glass, and providing a wooden door to keep the bees in darkness except when observations are being made.

— 16 —

Other Pets for Study

The study of our common domestic animals and beasts of burden is most interesting and valuable, but such study is, of course, best conducted in the home or in the field, as the case may be; but certain small mammals, kept in captivity, afford equally valuable opportunities for observation; and the care of such, handed over to the children, under proper supervision, does much to create an interest in animals, and to correct the all too common tendency to destroy living creatures or to treat them unkindly, while it must also lead to a more accurate conception of one's duty towards one's fellow creatures.

But in order that the above aims may be realized to the fullest possible extent it is absolutely essential that all pet creatures receive the utmost consideration as regards their health and comfort. Their homes should be large and roomy, and kept scrupulously clean, and the creatures themselves well fed and supplied with every necessity demanded by their individual habits. In order to obtain the greatest educational advantages from the study of the pets selected, all the children should, in turn, have the opportunity of taking an active part in their management; but the one or two scholars who, for the time being, are entrusted with the charge, should be allowed the sole responsibility with, of course, the guidance of a directing overseer.

While the larger pets are suitable only for outdoor hutches or cages, some of the very small ones are not at all objectionable in the classroom providing they are kept in a scrupulously clean condition. The former, however, should be housed in such a manner that they may be readily transferred to a sheltered spot on the approach of severe weather.

Wild animals that have been caught or trapped should never be caged; for although they may eventually become very tame, the early, and perhaps lengthy, period of timidity and constant efforts to escape can hardly have a beneficial influence on children who observe them. Yet we would not rigidly insist that every pet animal must necessarily be one that was born in captivity, for we have seen many instances where stray young animals, picked up in the open, have become almost immediately attached to their new keeper, and so well pleased with their new home and kind treatment that they have, from the very beginning of their new life, refused liberty.

The tamest and most fearless rabbit we have ever seen was, perhaps, the young one that, wounded and frightened, was rescued from the jaws of the voracious stoat, and transferred to a cozy hutch where it was well cared for by its rescuer; and the sauciest and most amusing of tame squirrels was probably the one captured as soon as it was strong enough to leave its nest, and similarly housed and treated.

Where several pet mammals are kept, it is advisable, if possible, that these be representatives of different groups, so that the observations may include those of a greater variety of habit and structure; but, unfortunately, the small British mammals suitable for the purpose are nearly all rodents, such as the mouse, dormouse, squirrel, and rabbit. There are several small British carnivorous mammals, but, even when perfectly tame, these species are always more or less objectionable as pets; and thus our observations of tame mammals are almost entirely restricted to the rodents, the larger domestic herbivorous beasts, and the carnivorous cat and dog.

As regards pet birds, other than the species which belong to the ordinary farmyard stock, many of the rules above laid down for mammals are equally applicable.

Wild birds that have been trapped must not be caged. The act is a cruel one; and, to any lover of Nature, the violent struggles for liberty, during which the imprisoned bird often wounds itself by its frantic efforts, present a most painful spectacle. Children should never be allowed to become familiar with such sights. Many wild species can be observed most easily in the open; and, in fact, it is only when they are under natural conditions that we can study the most interesting of their habits. It is an easy matter, too, to encourage many of our feathered friends to feed regularly in our own gardens and on our window sills, and to cause them to become so familiar with us that we may approach them very closely without producing any signs of fear.

Concerning the caging of wild birds, some would make exceptions in the case of several species, mostly foreign, that very readily adapt themselves to a life of captivity; but, generally speaking, we would restrict our observations of cage birds, at least for school study, to birds that have been reared in captivity. We have had, and seen, however, most interesting pets of our wild species that have been reared under exceptional circumstances, and which have become so tame and so thoroughly adapted to their homes and their treatment that they refused their liberty when the open cages were put outdoors.

Among these we may mention sparrows, thrushes and other birds that had fallen, when in a very young and helpless condition, from their nests, or which had been deprived of their parent or parents by the murderous gun of the sportsman or by some calamity, and then fed by hand until they were old enough to take care of themselves; also a chaffinch that was picked up in a starving condition on a severe winter's day, and warmed, fed, and comfortably housed. Feathered pets obtained under such circumstances become very strongly attached to their homes and their owners, and therefore give most favorable opportunities for study.

We can very strongly recommend the common ringed dove as a valuable pet for study, for this bird, when thoroughly familiar with its owner or owners, will allow itself to be very closely observed, and even to be very freely handled. This bird may be kept in a large cage outdoors throughout the greater part of the year, requiring shelter only when the weather is very cold.

We have had one for many years that would accept its food readily from the hand of any person, even a stranger, and feed its young without the slightest hesitation while closely surrounded by a crowd of observing children; that, when removed from its cage and placed on a table before a class of juveniles, would allow itself to be closely examined and handled, and permit its wings to be extended for the purpose of observing the joints and the arrangement of the feathers; and, while the object lesson was proceeding, this bird would never resent such interference, nor would it ever attempt to fly except so far as the shoulder or head of either the teacher or one of the children. Such tame creatures are exceedingly valuable for school studies, and such are easily to be procured as the result of kind and proper treatment.

— 17 —

The School Garden

The school garden might be made a very valuable aid to the study of Nature. In town schools, far removed from fields, woods and country lanes, the garden is a means by which interesting living things, that would otherwise be seldom seen, may be reared and studied by the children. In the case of village schools, already surrounded by wildlife in all its phases, it may be made the means of training the children in occupations which have a direct bearing on the industries of many of the inhabitants, thus assisting the scholars in understanding the principles underlying the work which some of them will be called upon to do in afterlife. And in all schools it should greatly help in training the children in habits of diligence, cleanliness, thrift and thoughtfulness.

It is advisable to allow as many children as possible to have a direct share in the work of the garden. The plots need not be large, and each one might be placed under the joint control of two or three scholars. The care of plants should commence in the very lowest classes. No child of school age is too young to watch the growth of a seedling, to observe the gradual development of a plant to maturity, and to learn the fundamental conditions under which such growth and development take place.

If no plot of ground can be spared for the youngest scholars, these may at least have a few window-boxes or flower-pots containing some plants or seeds that are tended by their tiny hands, and watched day by day by their admiring eyes.

All children, too, should be stimulated to secure, if possible, a small plot of the home garden, so that they may have the opportunity of putting into practice the principles learned at school, thus gaining more confidence in their labors, and securing additional material for study. And, whatever be the nature of the products grown, whether in the school garden or at home, every child concerned should be encouraged to keep a log-book in which to enter, with dates, a record of each operation performed and of the principal stages in the life histories of his plants or crops.

Furthermore, although it may be considered advisable at times to allow children, engaged on different plots, to grow the same product at the same time, either under similar or different conditions, in order that the results may be compared, yet it is well, as a rule, to introduce a considerable variety as regards the things cultivated. Each child will then have the opportunity of studying the varying productions of his schoolfellows' gardens; and if arrangements are made for a free circulation of the garden log-books among the members of the same class, so that each child is enabled to follow the records relating to the plot examined for the time being, the advantage to all will be enormous.

As regards country schools, we have already suggested that particular attention may profitably be given in the study of the produce raised in the locality for home consumption or for the market; but since the ability to raise such produce successfully is by no means an essential qualification of the country teacher, the advice and aid of a local gardener will often prove of great value.

With this aid, if such be necessary, let the children endeavor to cultivate some of the principal produce of the neighborhood; and, comparing the results with those obtained by the experienced hand, endeavor to find out the causes of any want of success. Let them compare the products of poor soil with those grown on rich land, and thus learn something of the nature and value of both artificial and natural manures. The senior children may be taught to bud and to graft fruit trees, and to watch the results of the pruning of both branches and roots; and, wherever they find results superior to those produced by their own efforts, they should be taught to search into the causes of the difference, and to strive for the best returns.

But although the production of marketable produce is a very useful training for the children of village schools, and is likely to be of much value to them in future years, yet, as far as the study of Nature is concerned, it is vastly inferior to the observations of wild plants and trees in their natural habitats. A child may learn to produce rare and choice flowers in a tidy and well-kept garden, but he will learn much more of Nature and her wonderful ways in a weedy bed, where all plants grow entirely undisturbed according to their natural habits. Here he will be able to see flowers which, though often small and inconspicuous, are natural and beautiful, to watch the varied and interesting habits of the plants, the struggle for existence, and the survival of those which are best fitted to maintain that struggle.

We very strongly recommend the study of our common weeds. Although, as a rule, we should insist on the thorough tidiness of garden beds, we lose much if we have not a patch of waste ground to observe. Let a small plot of ground grow wild, and even assist in increasing the wild character of the plot by scattering the seeds of wildflowers, and by setting roots of various herbs that have been dug out from banks, hedgerows and waste places. Here we shall have a curious mixture of erect, prostrate and decumbent plants, climbers and trailers, some producing accessory organs according to their present requirements, others creeping or twining or otherwise supporting themselves as best they can according to the exigencies of the situation.

Such a weed garden is most useful in the case of schools so situated that there are but few opportunities for observing plants in their natural habitats, for it provides a means of noting the habits of the plants grown, including the manner in which they struggle for the light when overcrowded and in which they strive to invade the surrounding territory, also how they fight for the most favorable conditions under which to perpetuate their species.

And it is not at all difficult to make arrangements for the purpose of growing many of the wildflowers in conformity with their habitats. Thus, we may prepare a bed in a shady corner, form a bank in a sunny spot, and throw up a stony heap, planting each with roots of wildflowers from corresponding situations, and supplying suitable supports as required by the climbing species.

Nor is it difficult to provide patches for the accommodation of plants which seem to be partial to particular soils in the neighborhood of the school, including, perhaps, a sandy bed, a bed of clay, a patch of chalky soil, a small bed of peat or other organic soil, a miniature marsh, and a small pond for aquatic plants.

A little garden pond suitable for the study of aquatic plants, and equally useful as a home for aquatic or amphibious animals, may be formed by digging a hole in the ground, beating the soil firmly all round to give it firmness, and then lining it with concrete; or, easier still, sink a large tub into the ground so that its edge is on a level with the surface of the soil. In either case, a layer of soil will be required in the pond to give the necessary hold for the roots of the plants introduced; and if the pond is always kept quite full, with rather frequent additions of water to cause a slight overflow, the surrounding soil will be kept sufficiently damp for the growth of some of the very interesting semi-aquatic plants.

The same arrangement will answer well for the miniature marsh, the only difference being that the pond or tub is practically filled with soil which is kept saturated with water.

Let the teacher, or rather the children superintended by the teacher, prepare such a garden as we have described, with its shady corner, sunny bank, little pond and miniature marsh, together with, if necessary, a few small patches of ground composed of special soils found in the vicinity of the school, and plant them all with wild plants collected during school excursions, and he will find it far more valuable, from a nature study point of view, than any well-kept garden filled with the choicest specimens of the florist's production. Such a garden will supply not only a wonderful variety of studies for the open air when the weather is fine, but will also yield abundance of material for closer examination in the schoolroom or at home.

And in addition to the various features mentioned above, the school garden may be made a valuable accessory to the study of various forms of animal life.

Not only will it form a natural home for all kinds of creeping things that invariably live in our gardens, but the flowers and plants will attract bees, flies and various other insects whose movements and habits are full of interest. If it is surrounded by a wall or perfectly-closed fence, it will permit the keeping of interesting little lizards and the harmless snake in a condition of semi-captivity; and, with the addition of the small pond, will form an admirable home for frogs, toads, and newts.

If the teacher is an expert in the management of bees, or, failing this, if the necessary assistance can be obtained from a neighbor, the establishment of an observation bee-hive will greatly increase the value of the garden.

Nesting box for birds (Fig. 189)

Furthermore, every endeavor should be made to encourage the birds of the neighborhood by fixing one or two little feeding tables and keeping them supplied with seeds, bread-crumbs, and other attractive foods. The little pond recommended will supply the birds with drinking water, and, with a little landing-stage sloping down into the water, will provide an excellent bathing-place for them.

Again, if the garden, or a portion of it, is planted with shrubs or small trees, it may form a suitable nesting place for some species, especially if it is furnished with suitable nesting boxes in places sufficiently concealed from the general view and traffic of the playground. The nesting boxes are very easily made, and our illustrations will supply all that is necessary for their construction, due regard being paid, of course, to the size of the birds which it is desired to encourage.

— 18 —

Our Garden Friends

We have already referred, on several occasions, to various creatures of the garden as providing suitable subjects for study; but, as a rule, we have not attempted to classify them, from the gardener's point of view, into friends and foes.

There seems to be a general tendency, on the part of those who rear flowers, vegetables and fruit, whether for pleasure or profit, to assume that all the creatures which inhabit the soil, creep on our plants, or pay passing visits to the garden, are necessarily injurious to our flowers and crops; and this assumption often leads to an indiscriminate slaughter of the creatures referred to, including many species which are valuable friends of those who destroy them.

In the present short chapter we propose to mention a few of our garden friends, and to present a few statements concerning their habits with a view of encouraging a closer observation and investigation.

First, then, as regards the various species of birds that frequent our gardens, both in town and in country, we must certainly look upon them as friends rather than as foes. It is true that some of them attack certain fruits, and occasionally devour buds and seedlings; but it must be admitted that these same species are generally greedy devourers of insects and other creatures which are very destructive to plants and trees; and that the small amount of damage they do is far outweighed by the valuable services they perform for us.

Those who have seen the sparrow attacking flowers and fruits, as it certainly will do at times, should also observe this busy little bird as it hops about among the garden plants, crops and trees, diligently searching for caterpillars and other marauders, and should endeavor to note what an enormous number of garden pests a single sparrow will destroy in a short time, especially during the breeding season when the young ones are being fed.

Tits, too, which sometimes attack buds during the winter months when they are unable to obtain much insect food, are very voracious devourers of various kinds of insects and grubs throughout the greater part of the year. In fact, there is strong reason for believing that tits are frequently accused of eating buds and fruits when they are in reality only searching out the grubs that attack these parts of our trees and bushes; and the same belief is probably equally applicable to the bullfinch.

Some of the seed-eating finches may be seen devouring the seeds of our garden plants, and picking up the newly-sown seeds that have not been properly covered by the soil; but it should be known that the principal food of these birds does not consist of garden seeds so much as those of wild plants, including many of the troublesome weeds that scatter their seeds far and wide to the great annoyance of the cultivator of the soil.

Then there are the soft-billed birds, such as the thrushes, blackbirds and starlings, which do us a good service by their wholesale slaughter of grubs, slugs, snails and other very destructive pests.

There is probably also a considerable amount of misunderstanding concerning earthworms. They are often accused of devouring seeds, and of eating young plants. This is not correct. Their food consists of decomposing organic matter that is mingled with the soil, and of the fallen leaves that are decaying on the surface, and they improve the richness of the soil by converting their organic food into a form more readily available for growing vegetation. Moreover, their burrows effectually drain the soil, rendering it warmer and preventing stagnation.

It is sometimes said that while earthworms do no harm in gardens and other cultivated ground, they are very injurious to plants grown in pots. Under the latter condition it is possible that the limited space for the movements of the worms makes it impossible for the creatures to burrow without interfering with the more delicate root fibers; and probably the limited supply of food contained in a pot of soil compels them to devour living structures which they would otherwise leave intact.

If earthworms are thus proved to be injurious to pot-plants, they can be easily brought to the surface by watering the soil with lime-water. The latter can be made by pouring a gallon or two of water on a piece of freshly-made quicklime, stirring the mixture well with a stick, and then pouring off the clear liquid after all sedimentary matter has settled to the bottom.

When earthworms are very abundant in well-kept grassland, the numerous castings thrown up on the surface may be somewhat objectionable; but, remembering that the worms do not injure the grass, but rather ventilate, drain and enrich the soil, it is better, and really less troublesome, to occasionally restore the surface of the soil than to attempt to exterminate the worms.

To accomplish the former it is only necessary to brush down the castings with a birch broom, and then roll the ground; but if it is considered necessary to reduce the number of earthworms, the safest plan is to go out with a lantern after dark, on an evening when the soil is moderately wet, pick off the worms seen on the surface, and either give them to the poultry or destroy them by dropping them into boiling water.

When earthworms are troublesome on gravel paths (and they never are where the paths are well kept and frequently rolled), they may be poisoned by means of a solution of mercuric chloride (corrosive sublimate); but the poisoned worms should always be buried, and never, of course, given to poultry. We do not, however, advocate the destruction of earthworms in any way. They seldom, if ever, prove troublesome in gardens where birds are not molested or where birds are encouraged.

Frogs and toads are very useful in gardens of all kinds, for they feed entirely on small living creatures, including many of the insects that are very destructive to our plants and crops. And centipedes, too, which are often destroyed under the impression that they attack plants, are carnivorous, subsisting on worms and insects which they hunt at night.

Although snails and slugs are, as a whole, very destructive vegetable feeders, yet there is one species of slug (*Testacella*) that is commonly spoken of as a gardener's friend because it feeds on earthworms. This slug is not very common, and is really an introduced species as far as Britain is concerned. It is of a light color, with a small ear-like shell on the posterior part of the body. It is subterranean in its habits, coming to the surface of the soil only at night.

We are very doubtful as to the usefulness of this slug in the garden. If, as we have said, earthworms are really garden friends, then we can hardly regard this worm-eating slug as a valuable inhabitant of cultivated ground, unless it be to reduce the number of worms in places where they are so numerous as to be objectionable.

The Testacella (Fig. 190)

The majority of the creatures that commit serious ravages on our garden produce belong to the insect world, and yet in this group we are able to discover several species that are to be regarded as valuable friends. Many of these belong to the beetle order, including lady-birds, the devil's coach-horse, and various species of ground beetles.

Lady-birds and their larvae or grubs feed on aphides (green blight or green fly), and they devour such vast numbers of these pests that they are often regarded as among the greatest benefactors of the agriculturist, and have been credited with the saving of a good crop that would otherwise have been entirely destroyed. Lady-birds may be seen creeping over garden plants, searching out the aphides, devouring them, and depositing clusters of eggs in their midst. The larvae that issue from these eggs, though apparently very sluggish and inert, continue the work of their more active parents, devouring the aphides in very large numbers. When the larvae are fully grown, they suspend themselves, head downwards, from a leaf, and change to quiescent pupae from which the next generation of lady-birds emerge.

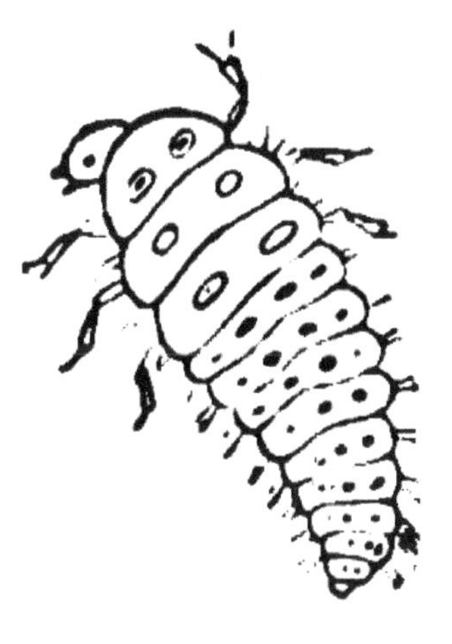

Larva of the Lady-bird (Fig. 191)

The devil's coach-horse or cocktail beetle is so common that it must be known, though not, perhaps, by name, to everyone who has had anything to do with garden operations, and our illustration, which shows characteristic attitudes of the beetle, will be all that is required for its identification.

Many other insects of this group, including several species of ground-beetles and tiger beetles, are greedy devourers of various kinds of grubs, and are therefore exceedingly useful in the garden. These beetles are equally valuable both in their larval or grub condition and in the perfect state. In the former state they may generally be recognized by their whitish, segmented bodies, with three pairs of legs attached to the three segments next the head.

Ichneumon flies are also to be included among the most valuable of our garden friends. These flies belong to the same order of insects as ants, bees and wasps, having two pairs of wings, the hindmost much smaller than the front ones, and the two wings on each side of the body locked together by means of a row of little hooks during flight. They vary very considerably in size, some being as large as wasps, which they often much resemble, while others are so small that a hand lens is required to identify them.

These flies are the parasites of other insects. They may be seen flying among the plants in gardens, searching for caterpillars, on the bodies of which they deposit their eggs; or, piercing the bodies of the caterpillars with their *ovipositors*, lay their eggs inside them.

The grubs that issue from these eggs live within the caterpillar, devouring its body as it grows; and, about the time when the caterpillar is fully grown, the grubs, also fully grown, eat their way out, change to pupae, and eventually to perfect flies. The caterpillar itself is generally left in such a weak and lean condition that it dies instead of developing into a butterfly or a moth.

The Devil's Coach-Horse (Fig. 192)

A very pretty insect, known as the lace-wing fly, is also a useful creature in the garden. This fly is easily recognized by its leaf-green body, large gauze-like wings that exhibit transient hues of green and pink, long antennae, bright golden eyes, and—its unpleasant odor. Its larva feeds on aphides or 'green fly,' of which it devours large numbers.

The eggs of the lace-wing fly are very interesting objects. They are white, and are always mounted on the ends of slender threads attached to a leaf or stem. The thread is formed of a sticky liquid that is drawn from the body of the fly and which hardens immediately on exposure to air.

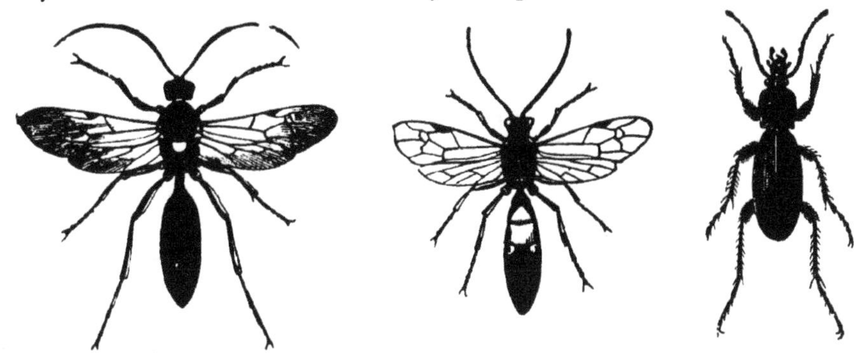

Ichneumon flies (Fig. 194) and the Violet Ground Beetle (Fig. 193)

Spiders are sometimes included among our garden friends, but, although it would be difficult to prove that they do much harm, yet it appears very doubtful that they can be of much use. They certainly entrap and devour large numbers of flies, but it must be remembered that these flies include both friends and foes. The grubs of some of them attack our root crops, while those of others are injurious in various other ways.

The Lace-wing Fly and its eggs (Fig. 195)

On the other hand, the larva of some of the flies are, like those of the ichneumons, parasites that feed within the bodies of destructive caterpillars, while those of other species devour decomposing matter that would otherwise putrefy and poison the air, and thus are valuable scavengers. Furthermore, some of the flies devoured by spiders are species which visit flowers, and therefore aid considerably in transferring the pollen that is necessary for fertilization. Those who have observed the habits of spiders, and noted the habits of the flies that have been seen in the webs, must decide for themselves whether they will claim the spiders as their friends or pronounce them foes.

— 19 —

Garden Foes

Garden foes are far more numerous than garden friends. From a nature study point of view it is most unfortunate that this should be the case; for although the study of garden pests is quite as interesting and instructive as that of the friends, we are often under the painful necessity of pointing out to our children various creatures which must be destroyed unless we are willing to submit to their ravages, with the possibility of the total loss of some of our valued plants or crops.

One of the main objects of a course of nature study is to put the child in sympathy with Nature in all its phases—to create such an interest in all kinds of living things as will lead the child to see their beauty and their wonderful habits, and thus teach it to admire rather than to destroy. Yet, on the other hand, if we are to train children in the pleasant occupation of rearing flowering plants and raising crops and fruit, we are bound to encourage them to distinguish between friends and foes, and even to wage a war of extermination on the latter.

It will be seen at once that this is a matter of the greatest importance in agricultural districts where a large proportion of the children will eventually be employed in the raising of garden and field produce for home consumption or for the market. So, while we do our best to point out the beauties of all forms of life, we have, at the same time, to encourage such observations as will lead to the discovery of the habits of objectionable species, and to the best means of reducing their numbers.

Since the number of garden pests is so great, we shall necessarily have to be very brief in our remarks concerning them, and even have to omit entirely many that must, sooner or later, claim the attention of every possessor of a garden. Our main object will be to enable the reader to identify his foes, and to give such hints as may lead to a discovery of their habits and to the ordinary methods of reducing their ravages.

Most of the foes belong to the insect world, so we will regard these first, grouping them according to their structure and habits rather than to the nature of the plants which they attack.

Caterpillars are particularly destructive. They are the grubs or larvae of butterflies and moths which, in the perfect or winged condition, do no harm except to deposit the eggs that are to produce a new generation of grubs.

When an unknown caterpillar is found feeding on one of the plants or trees in the garden, put it in a larva cage such as is described on page 217, feed it on its proper food until it changes to the pupa or chrysalis, and keep the latter until the emergence of the perfect insect takes place. By this means one becomes acquainted not only with the habits and metamorphosis of the insect concerned, itself a most interesting study, but also learns to associate the objectionable grub with its final condition. This latter end being accomplished, much may be done towards the extermination of the pest by destroying the butterfly or moth that would otherwise deposit a large number of eggs and so give rise to as many grubs which would continue the destructive work of their predecessors.

The following are a few brief notes that will assist in the identification of the commonest of our destructive caterpillars and of the butterflies and moths which give rise to, and produce them:

The Large White or Cabbage Butterfly caterpillars (Fig. 41) are of a greenish color, with three yellow longitudinal stripes, and several black projecting spots each bearing a short hair. They do an enormous amount of damage among cabbages and nasturtiums during spring, and again in the summer, for there are two distinct broods in the year. When fully grown they change to angular chrysalides, of a bluish white color with black spots, and these may be seen in sheltered positions on garden walls and fences, secured by a fine but strong silken cord.

The chrysalides of the first brood should be looked for about June, and of the second from September to the following March; and the butterflies appear during April and May, also during July and August. The butterfly may be identified by the aid of our illustration, and the male may be distinguished from the female by the absence of black marks on the front wings with the exception of those at the tips. The eggs, which resemble very minute yellowish vases, are to be seen on the food plants in May and July.

The caterpillars are attacked by small ichneumon flies (page 234) the larvae of which, after living as internal parasites, emerge from the unfortunate hosts, by the dead or dying bodies of which they construct little yellow cocoons previous to changing to pupae.

Currant Clearwing Moth (Fig. 196)

The Ghost Moth (Fig. 197)

The Small White Butterfly is also a troublesome species, its larva feeding on cabbages, rape, horseradish and other cruciferous plants, as well as on the nasturtiums of our flower gardens. It is a much smaller species, as will be seen from our illustration, but otherwise similar both in appearance and habits, and its life history also corresponds with that of the last species. The caterpillar is of a glaucous green color, with a yellow stripe and yellow spots, and the whole of its body is covered with very short hairs. The chrysalis is grey, brownish or greenish, frequently marked with spots of a darker color.

Currant and gooseberry bushes are often seriously damaged by the caterpillar of the little Currant Clearwing Moth, which burrows into the twigs, and feeds on the pith till it is fully grown. It then changes to a chrysalis within the stem, and the perfect insect emerges through a hole in the side of a shoot. The moths may be seen on the wing in June, and the grubs occupy the stems throughout the autumn, winter and spring. The pithless twigs infested by the grubs should be cut off and burnt.

A pale yellow larva, with a brown head and a brown, horny scale on the front of the next segment of the body, may often be seen feeding on roots throughout the winter. This is the caterpillar of the Ghost Moth or Ghost Swift; and although it subsists largely on the roots of weeds, it is frequently very destructive to those of hops and other crops. The female moth, which has yellow fore wings marked with wavy reddish lines, and smoke-colored hind wings, is shown in the illustration. The wings of the male are white and silky, with very narrow brownish margins.

The Goat Moth and its larva (Fig. 198 & 199)

Apple and other trees are frequently drilled with holes at the mouths of which small fragments of wood resembling sawdust will be seen, and the trees are sometimes riddled with these perforations to such an extent that they are weakened or killed. This is the work of the larvae of a large moth called the Goat Moth. They are of a reddish brown color above, and flesh-colored beneath, with black heads; and they feed on the solid wood. They are not fully grown till about three years old, and then they change to a brown chrysalis in a cocoon constructed by binding together the bitten fragments of wood with silk. The moth emerges in June or July.

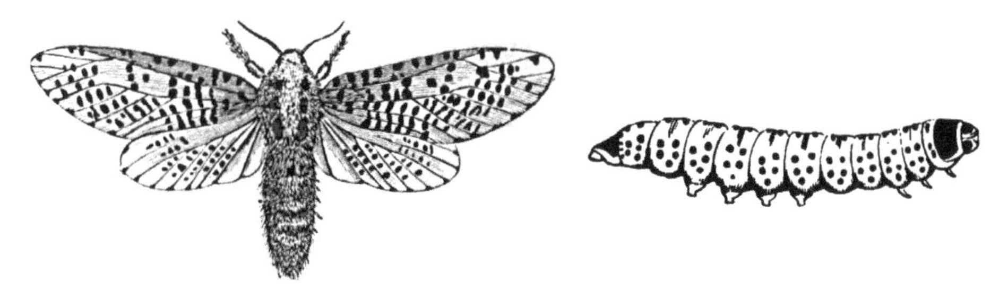

The Goat Moth and its larva (Fig. 200 & 201)

Similar damage is done to fruit and other trees by the caterpillar of the Leopard Moth. This grub is yellowish, with glossy black spots and a black scale behind the head; and the semi-transparent wings of the moth are white with blue-black spots. Branches of trees infected with this or the last species should be cut and burnt during the winter to destroy the grubs.

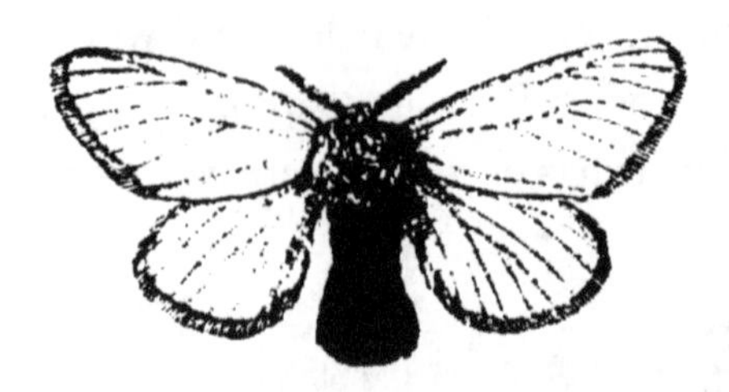

The Brown-tail Moth (Fig. 202)

A pretty moth called the Brown-tail, with white silky wings, lays its golden eggs on the underside of the leaves of the apple and other trees. These eggs give rise to caterpillars which devour the leaves, living at first under the cover of a common silken tent, but afterwards separating. The caterpillars are black, with brownish hairs, two red lines along the back, and a broken white line on each side. They hibernate through the winter, and are not fully grown till the following summer. The moth may be seen on the wing in August and September.

The male and female Vapourer Moth and their caterpillar (Fig. 203, 204, & 205)

Fruit trees are also seriously damaged by the ravages of the caterpillar of the Vapourer Moth. This caterpillar, which is variously colored with brown, grey and pink, and ornamented by peculiar tufts of hairs, may be recognized by our illustration. When full grown, in the summer, it spins a silken cocoon on the bark of a tree or on a fence, and changes to a hairy chrysalis from which the perfect insect emerges in August or September. The male moth is of a chestnut brown color, and, unlike moths in general, flies about in the sunshine. The female is wingless, and soon after she emerges from the chrysalis she lays her eggs on the outside of the cocoon. The best way to reduce the numbers of this insect is to search for the egg-covered cocoons during the winter months and destroy them.

Several of our common trees, including fruit trees, are attacked by the caterpillars of the Buff-tip Moth. These caterpillars are gregarious in their habits, and completely strip a branch of its leaves before they leave it for another. They are of a dull yellow color, hairy, and marked by several broken, black lines. The moth is so called on account of the buff-colored patches at the tips of the front wings. The young caterpillars should be searched for about the end of June, at which time a large number may be seen crowded together on a single leaf. As they grow larger they do not keep so close together, and are then best secured by beating the branches to cause them to fall to the ground, or by picking them from the leaves singly.

The Buff-tip Moth and its caterpillar (Fig. 206 & 207)

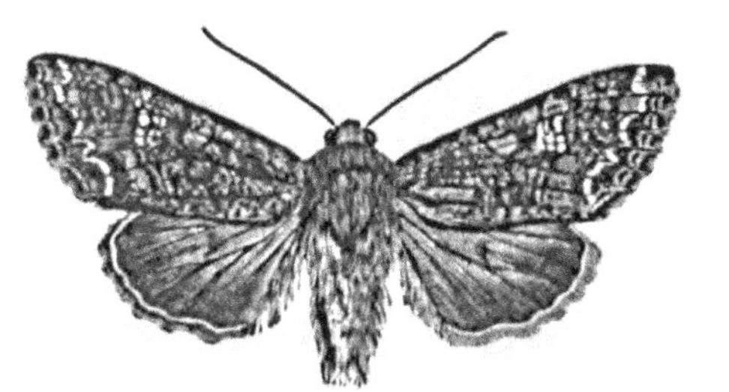

The Cabbage Moth (Fig. 208)

The Turnip Moth (Fig. 209)

One of the most troublesome pests of the vegetable garden is the caterpillar of the Cabbage Moth. This grub, which is of a dark grey color, with a darker line on the back and a lighter one on each side, burrows into the hearts of cabbages. It also attacks many other plants both of the vegetable and of the flower garden, as well as various weeds. It changes to a chrysalis beneath the surface of the soil in the autumn, and the moth emerges about June.

The front wings of the moth are of a dull brownish-grey color, mottled with darker tints, and marked by an irregular, lighter line parallel with the hind margin; and the hind wings are smoke-color. The caterpillars should be removed by hand and destroyed, and their ravages may be lessened by means of a liberal supply of strong soapy water. A frequent hoeing of the ground in autumn and winter will expose or destroy the chrysalides.

The Turnip Moth is somewhat similar to the last, but its fore wings are brown, and the hind ones very much lighter. Its caterpillar is greyish or greenish, with a light line along the back, a light brown line on each side of this, and black spots between these lines. It feeds on turnips and the roots of other crops, often doing much damage. The caterpillar generally lives throughout the winter, during which season it continues its destructive work.

We give an illustration of the Large Yellow Underwing Moth and its caterpillar. The former is easily known by the bright yellow hind wings with a broad, black border. The latter, which is of a dull yellowish or greenish color, feeds on the roots of various plants throughout the winter, and, ascending from the soil in the spring, commences to attack the leaves and stems.

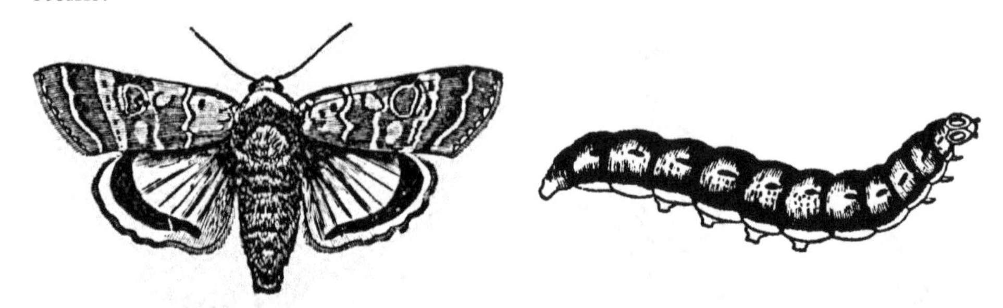

The large Yellow Underwing Moth and its caterpillar (Fig. 210)

Those who cultivate currants and gooseberries are sure to become acquainted with the caterpillars of the Currant or Magpie Moth, which are often so abundant that they completely strip the bushes of their leaves and buds, thus totally preventing the development of flowers and fruits. These caterpillars are colored with a creamy white, heavily dotted and blotched with black. They are 'looper' caterpillars; that is, they creep by alternately looping and extending their bodies. They first appear from the eggs in the autumn; and, while still small, hibernate for the winter in spun leaves either on the bush or on the ground. As soon as the warmer weather comes they commence feeding on the buds, young leaves and flowers of currant and gooseberry bushes; and, if left undisturbed, are full grown in June, when they change to black chrysalides with yellow bands.

The moth, which has creamy white wings with black blotches, may be seen, from the end of June to August, flying in the daytime. The young caterpillars should be destroyed, as far as possible, in the autumn, and the ground well hoed beneath the bushes. A careful search is also necessary in the spring in order to secure those not previously seen. It is interesting to note that the caterpillars of the currant moth are not attacked by birds, frogs and other insect-eating creatures on account of their objectionable taste, but they are the victims of the parasitic grubs of ichneumon and other flies.

The Currant Moth (Fig. 211) *The Codlin Moth (Fig. 212)*

A little white grub is commonly found inside an apple or a pear. This is the caterpillar of the Codlin Moth—a small moth measuring less than an inch from tip to tip when its wings are expanded. One egg only is laid on each young fruit, and the caterpillar produced from it burrows into the heart of the latter. When full grown it lets itself to the ground on a silken thread, creeps to a neighboring tree, ascends the trunk, and changes to a chrysalis within a silken cocoon in a crevice of the bark.

From this chrysalis the moth emerges in June or July. In order to reduce this troublesome pest all fallen fruit that shows evidence of being worm-eaten should be destroyed by burning, as well as all dead rubbish that may have accumulated beneath the trees. In order to catch the caterpillars as they ascend the trunks to pupate, tie wide strips of paper around the base of the tree, and smear them thickly with cart-grease. The trees should also be sprayed in the spring with an insecticide, to kill the grubs that may have entered the young fruit.

The Small Ermine Moth is very destructive to fruit trees and various other trees and shrubs. It is a very small moth, measuring not much more than half an inch from tip to tip, its white fore wings dotted with black. The little caterpillars of this species live together under the protection of a silken web, and often entirely strip branches of their foliage. All twigs supporting their web should be cut off and destroyed before the caterpillars within are fully grown.

Leaves rolled & mined by small caterpillars (Fig. 213)

The only other insects of this group to which we can refer are the leaf-mining and leaf-rolling caterpillars, both of which are the larvae of small moths. The former burrow into leaves, feeding as they go, but always leaving the epidermis of the foliage intact. When they are fully grown they change to little chrysalides within the leaf, and the moth emerges shortly after. The latter roll up leaves, binding them with their silk, thus making for themselves a comfortable home of the same material that serves them for food.

These little pests are not nearly so destructive as many of the species before mentioned, but they often do considerable damage. All leaves that are rolled or mined should be destroyed, with their occupants, before the latter are fully grown. By this means we prevent the appearance of the moths that would deposit large numbers of eggs for the production of a new generation of grubs.

The June Bug and its grub (Fig. 214 & 215)

A few at least of our garden pests belong to the beetle tribe of insects, and first among these we mention the Cockchafer Beetle or 'June Bug,' the large, fat, white grub of which devours the roots of plants in enormous quantities, and, after continuing this destructive work for a period of about three years, changes to the perfect state in which it does further damage by devouring the leaves of trees.

The grubs are often very ruinous to potato and other crops, and should always be destroyed when found; but it is almost exterminate them, since they occur in prodigious numbers in pastures, where they eat the roots of grasses. Our friends the rooks do much to keep down this troublesome pest, for they drag large numbers of the grubs out of the ground, and either devour them on the spot or take them home to their young.

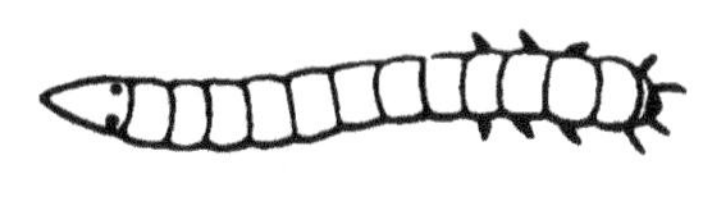

The Wireworm (Fig. 216)

A great enemy to the agriculturist is the grub known as the wireworm, for it commits terrible ravages among plants, attacking the root-crops as well as the roots of plants of all kinds. It is not a worm, as may be known by the presence of its three pairs of legs, but the larva of a beetle known as the Skipjack or Click Beetle. Large numbers of the grubs are devoured by rooks, starlings, and other birds. If they are discovered among the roots of garden plants they may be trapped by placing pieces of potato on the ground. Numbers of them will burrow into the potatoes, which may then be collected up and destroyed.

There is a group of little beetles known as the Weevils, the members of which are generally to be distinguished readily by the presence of a beak or snout. Their antennae or feelers are situated at or near the end of this beak, and are generally elbowed or sharply bent.

Many of these beetles are very destructive to trees and plants, some of them confining their ravages to one particular species, while others are not so restricted as to their food.

Among them we may mention the Pea Weevils which are very injurious to crops of peas, clovers, and other leguminous plants. They are very small beetles, which bite away the leaves and tender shoots of the plants, while, in the larval condition, as footless grubs, they devour the roots. Spraying the plants with soapy water containing a little paraffin will render them distasteful to these insects; and a dressing of soot and lime on the soil will do much towards the destruction of the grubs.

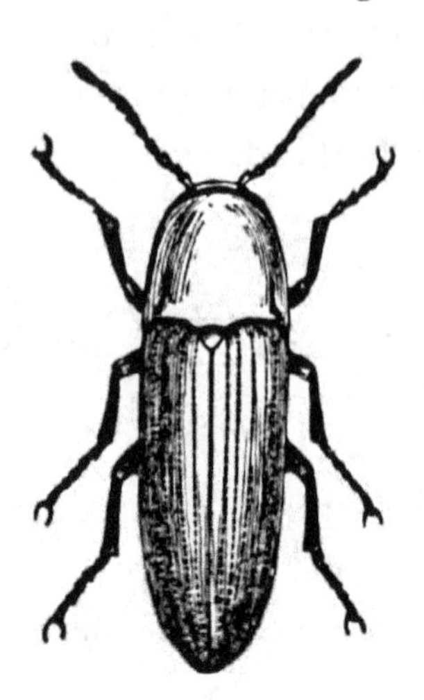

The Click Beetle (Fig. 217)

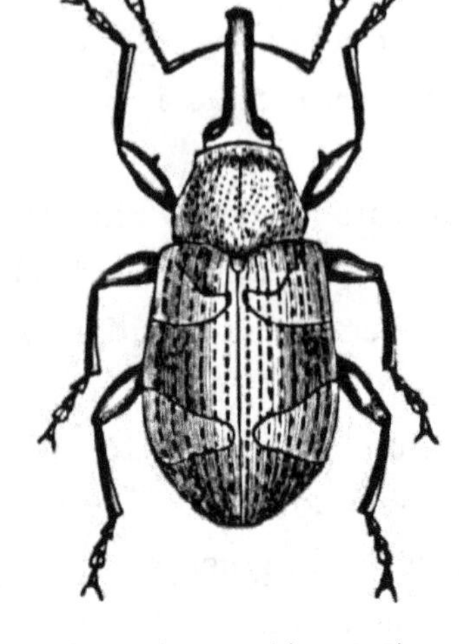

The Apple Weevil (Fig. 218)

Other weevils, some of them less than a tenth of an inch in length, burrow into fruit and other trees, and lay their eggs beneath the bark. The grubs which are hatched from these eggs eat their way into the inner bark, thus interfering with the circulation of the sap, and often causing the death of the tree. When a tree is once infested with these pests there seems to be no way of destroying the intruders except by removing and burning the branches that have been attacked, or, if necessary, by burning the whole tree.

The Apple Blossom Weevil does a deal of injury to the apple and pear flowers in early spring. The perfect insect, which has been hiding in the crevices of the bark throughout the winter, ascends the tree about the end of March, bores a hole in the flower buds, and deposits an egg in each. The young grubs which are produced from these eggs devour the essential parts of the unopened flower, thus preventing the development of fruit. This beetle, and other similar marauders of our fruit trees, may be reduced by spraying the trees with an insecticide. A liberal spraying of the bark very early in the year will kill the perfect insects before they rise to the buds for the purpose of depositing their eggs.

Saw-flies are somewhat wasp-like in general appearance, and belong to the same order of insects as wasps, but they are generally smaller than the latter, and have not such a narrow, thread-like waist. The female insect is provided with a pair of small saws at the tip of her abdomen, and by means of these she cuts grooves in plants, and deposits an egg in each. The larvae to which these give rise are much like caterpillars (the grubs of butterflies and moths), but they have a larger number of appendages. The caterpillar never has more than sixteen of these, including the true legs and the pro-legs, while the larvae of the saw-flies have from eighteen to twenty-two.

The Rose Saw-fly (Fig. 219)

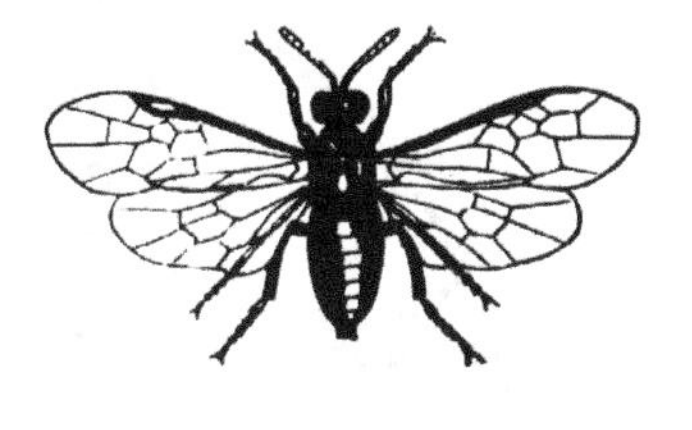

The Turnip Saw-fly (Fig. 220)

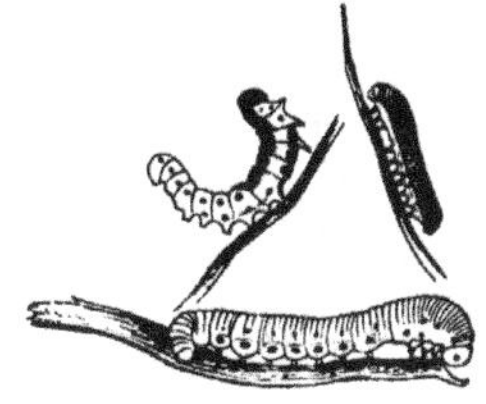

Sawfly larva (Fig. 221)

These larvae do considerable damage to some of our crops, and among them we may mention the Rose Saw-fly, which attacks rose bushes. Its larva may often be seen on the bushes, where it rests with the hindmost portion of the body bent upward. Another species—the Turnip Saw-fly—is often very abundant in turnip fields, where the larvae devour the leaves of the turnips, thus preventing the development of the roots.

The larvae should always be destroyed when found; but, as regards the last species, it is said that there is no remedy so effectual as that of turning a large number of ducks into the turnip field, for these birds greedily devour the grubs, and become well fattened thereby.

Since the larvae descend into the soil, when fully grown, to change to the pupal condition, and remain in this state until the following spring, a thorough hoeing of the ground in the autumn, and again early in the spring, will probably destroy considerable numbers.

A third saw-fly, known as the Apple Saw-fly, is very destructive in fruit gardens and orchards. The larvae of this species burrow into apples, not making a tunnel as does the caterpillar of the Codlin Moth, but eating out a cavity in the center of the fruit. All fallen fruit attacked by these pests should be burnt. Quicklime, spread on the ground in early spring, will kill the pupae in the soil; and a hoeing of the soil beneath the trees will aid in attaining the same end.

Curious swellings of various shapes are often to be seen on our plants, bushes and trees, generally attached to the veins or the stalks of leaves, to young twigs, or even to the roots. These swellings, known as galls, are produced by the agency of Gall-flies, of which there are many species, each one giving rise to a different kind of gall on its own favorite plant or tree.

The gall-fly—a four-winged fly belonging to the same order of insects as bees and wasps—pierces the vegetable structure with its ovipositor (egg-depositing organ), at the same time introducing an egg together with a tiny drop of an irritating fluid. The effect of the latter is to cause the swelling or gall which provides both food and home for the grub when it hatches out. The grub remains within the gall until it is fully grown and has undergone its metamorphosis, and the perfect insect then gnaws its way out to the air. Thus, if no opening exists in the gall, we may know that the developing insect is still enclosed.

A Gall-fly and Galls (Fig. 222)

It is perhaps fortunate that the gall-flies attack wild plants and forest trees more than those of our gardens and orchards; but, in any case, when it is desired to reduce the numbers of these pests, the galls should be picked off before the fly emerges, and then burnt. The most familiar examples of galls are the really pretty, moss-like bedeguar galls so commonly seen on rose trees, particularly on wild roses; and the oak-apple and currant galls that are even more common on oak trees.

Many of the insect foes that infest our gardens are destructive only in the grub stage, though, of course, we hold the perfect insects equally obnoxious since they are responsible for the production of the grubs. Thus the well-known Daddy-long-legs, that is so abundant on grassland, flying about among the blades during the summer, is not in itself destructive, but its grubs do an immense amount of mischief by devouring the roots of grasses, and also by gnawing away at the stems of some of our vegetable crops and flowering plants.

This grub is popularly known as the Leather-jacket on account of its very tough skin. The perfect insect has only two wings, and belongs to the same order of insects as the house-fly and the gnat.

Earwigs are a great annoyance to the cultivators of flowers, for they climb up the plants at night and bite away the petals of flowers, especially dahlias and carnations. They hide during the daytime in almost any sheltered place they can find, and it is easy to trap them by providing a suitable hiding place on or near the plants they infest. Small flower-pots, loosely packed with moss or hay, and placed on the tops of the sticks used to support the plants, form very effectual traps. Cones of stiff paper, or paper tubes plugged at one end, similarly packed, are equally good. The earwigs that have availed themselves of the shelter thus provided should be shaken out into a pail of water in the morning, or at any time in the day.

Notwithstanding the objectionable habit of earwigs referred to, they are certainly most interesting insects, and a study of their life histories and movements will well repay the time spent in their observation. It is seldom one sees earwigs on the wing, but they fly freely at night. Their wings are exceedingly delicate in structure, and are beautifully folded, when not in use, beneath the two short wing-covers seen on the top of the body, just behind the head. The forceps at the tip of the abdomen are employed in folding the wings when the insects alight.

Earwigs do not pass through three distinct stages, as do most of the other insects we have mentioned. The young ones are very similar to the adults except that their wings are not developed.

We must now say a few words concerning the Aphides or Plantlice, also commonly known as the Blight, which are among the most troublesome of our garden pests. These are small insects, usually of a green color, and therefore commonly spoken of as green-fly. They are provided with slender beaks by means of which they pierce the tender twigs, leaves and buds of plants and trees, and suck out the sap. They also multiply very rapidly, and are frequently so numerous that twigs are completely covered by them and so effectually drained of their sap that all future development is stopped.

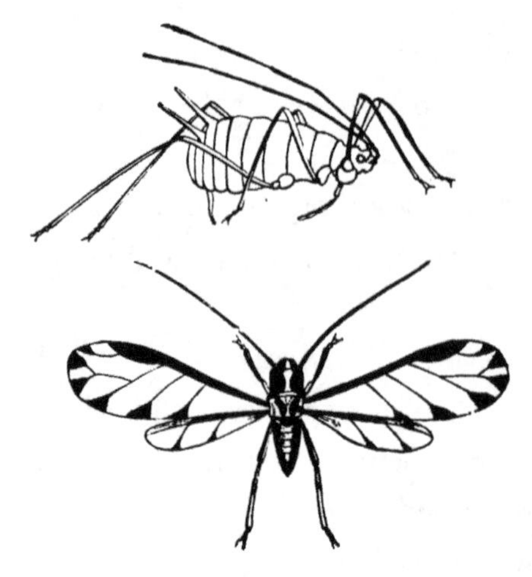

The Lime-tree Aphis male (winged) and female (wingless)(Fig. 223)

It will be noticed that the aphides of early summer are wingless, while some of those which appear later in the year are provided with two pairs of delicate wings. Some species have also a pair of appendages from which exude a liquid, tiny drops being discernible at the tips of these appendages, which are really tubes. This fluid is often produced in such quantities where the aphides are numerous that it drops to the ground, where it is greedily devoured by ants. Ants also seek out the aphides for the purpose of obtaining the fluid, and sometimes even convey the creatures to their nests and take care of them for the sake of the food thus derived from them.

We have already mentioned the fact that lady-birds and their larvae are great enemies to the aphides, and that, on this account, the lady-birds and their grubs should never be destroyed. It will not do, however, to trust to these friends alone for the removal of so destructive and abundant a pest. Whenever aphides are seen they should be brushed into a vessel of water, and the plants or trees so affected should then be sprayed with a strong solution of soft soap to prevent further intrusion.

One species of aphis, known as the American Blight, is a woolly insect that is commonly seen on apple and other trees. The body is covered with a white down, and this renders the insects very conspicuous when they are clustered in the crevices of the bark. The downy covering provides a means by which these aphides are readily blown about by the wind. Twigs infested with this blight should be sprayed as recommended above; and when the aphides appear on the trunks of trees a liberal application of freshly-made limewash, well brushed into the crevices of the bark, will do much to reduce their numbers.

We frequently see little masses of a frothy substance on our flowering plants and on the shoots of shrubs and trees. On pushing this aside we find that it enclosed a little white grub with conspicuous black eyes.

This grub is the larva of an insect known as the Froghopper or Cuckoo-spit. It is provided with a sharp beak with which it pierces the epidermis of the plant or shoot and sucks out the sap.

Although this creature does not do nearly so much damage to the vegetation as the aphides, yet the frothy matter that it produces to hide itself from its enemies looks very unsightly. The larvae should be removed with a small camel-hair brush, and then destroyed. When they occur in very large numbers they may be washed away by the use of a syringe or by means of the garden hose, but it is probable that many of them will again ascend the plants and continue their objectionable work if they are not killed.

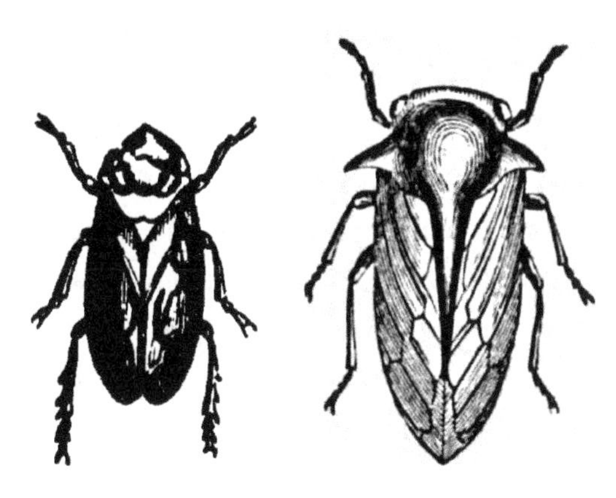

Two species of Frog-hoppers (Fig. 224)

These larvae develop into winged insects which may be seen resting on the leaves of plants, and which take prodigious leaps into the air, when disturbed, by means of their powerful hind legs. The wings are four in number, and all are membranous and similar.

There are many species of Plant-bugs, with habits similar to those of the cuckoo-spit and aphides, inasmuch as they pierce plants with their beaks and suck the sap. It is doubtful whether they really do a great deal of damage, but when it is desired to reduce their numbers perhaps the most effectual plan is to shake them from the plants by giving the latter a smart tap, holding a muslin net beneath to catch the insects as they fall.

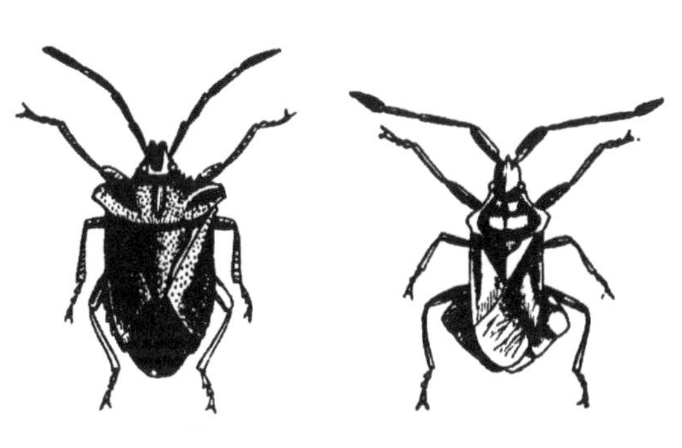

Plant-bugs (Fig. 225)

Up to the present all the garden foes mentioned belong to the insect world, and undoubtedly this division of the animal kingdom includes by far the greater number of these pests. A few, however, known as Mites, belong to the spider class, and are distinguished from most insects in that they do not undergo metamorphosis.

One of these—the Currant Mite—infests the buds of currant bushes, particularly those of the black currant. This mite seems to creep over the bushes during early summer, and later in the season to seek shelter within the young buds and in the bark, where it sucks the sap. Buds infested with these mites become larger and round, and all such buds should be removed and destroyed. It sometimes happens that practically every bud of a bush is attacked by them, in which case it is advisable to dig out the bush and burn it completely. Spraying the bushes with water in which is stirred a mixture of lime and flowers of sulfur will often prevent the attack of mites.

Animal garden pests are so numerous that we are unable, in the space at our disposal, to deal with even all the very common ones, and we shall conclude our list with a brief note on the snails and slugs.

These are nocturnal creatures, hiding in the daytime in places where they are sheltered from the direct rays of the sun, and committing their ravages on tender leaves during the night. The outside leaves of the cabbage, spread on the ground in spots where slugs and snails are most numerous, will not only supply them with food, and thus tend to prevent them from attacking the plants, but will also provide them with attractive shelters from which they may be removed at leisure. Lime or soot, or a mixture of both, spread on the soil around the roots of the plants that are most frequently attacked or that are most highly valued, will prevent the approach of these foes during the night.

Some of our garden foes, including species which do a considerable amount of damage, are parasitic plants (*fungi*) of a low type, often so minute that they can be seen only with the aid of a compound microscope.

Threadlike molds are very destructive agents in the production of disease, and are the cause of the potato murrain. White molds, known as mildews, affect the leaves of plants and trees, and also fruits. Other fungi attack the flowers of wheat, oats and barley, producing that black, powdery mass known as 'smut.'

Other minute fungi produce diseases known as rot, and these may attack roots, stems, leaves or fruit. When the roots of trees are affected the leaves generally lose their fresh color and wither; and in this case it is advisable to expose as much as possible of the roots and dust them liberally with powdered sulfur. If the leaves or fruit be themselves attacked, they should be gathered and burnt; and the spreading of the disease may be prevented by spraying the trees with a solution of sulfate of iron. Watering the ground with this solution will also help to destroy any spores of the fungi that may be on or in the soil.

It is impossible to deal here with all the various diseases of plants and trees that are caused by minute forms of vegetable life; but we strongly recommend all teachers of country schools who desire to train their children in the cultivation of vegetables and fruit to procure the valuable leaflets issued by the Board of Agriculture and Fisheries. These leaflets are supplied gratis on application to the Secretary to the Board, Whitehall Place, London.

— 20 —

Nature Lantern Slides

This short chapter is to consist of a few hints concerning the production of simple lantern slides for the illustration of nature lessons; but before giving the necessary instructions we would repeat our views as regards the proper use of such illustrations.

In the first place, the slides should never be used for the purpose of showing features which the children can observe for themselves direct from Nature. They should never be used as substitutes for natural objects or phenomena, but only to assist the teacher in directing the children's observation of the objects placed before them, to illustrate points of structure which the children cannot see for themselves, or to represent scenes and phenomena that the children have no opportunity of witnessing.

The best slides for the above purposes are undoubtedly photographs direct from Nature, and such, in great variety, are to be hired, at small cost, from numerous dealers in optical appliances. Many teachers, however, are themselves more or less expert in the production of photographic representations of Nature, and these are placed at a great advantage as regards the illustration of their nature lessons. But we shall assume no such ability, and deal only with a few simple means of producing suitable slides that are within the reach of all.

1. Pen-and-ink Sketches on Plain Glass

Procure some pieces of plain glass, thin, and free from flaws, cut to exactly 3 ¼ in. square—the standard size for lantern slides. Such glasses may be obtained from any glazier, but it is far better, as a rule, to purchase those sold by opticians for lantern slide work. These latter are cut from a glass of suitable quality, are very thin, and always of exactly the right size. The sketches required should be made with a fine pen, using either indian ink or Stephens' ebony stain. Of course the glasses should be thoroughly cleaned before using; and, if it is found that the ink or stain does not flow easily with the strokes of the pen, add a few drops of oxgall to the former. If the sketch desired is to be reproduced from a book illustration without any alteration in size, it may be traced.

When the ink or stain is quite dry, cover it with a second glass, and bind the two together by means of strips of paper, about half an inch wide, fastened with ordinary paste; or by means of the strips, ready coated with an adhesive, sold expressly for the purpose.

If it is desired to tint the sketch, or any portion of it, transparent colors only should be used, the best being the varnish colors made especially for this work.

A second glass for the protection of the sketch is not absolutely necessary, but, of course, the latter is easily damaged if not protected in some way. Dispensing with the 'cover glass,' a thin layer of transparent varnish is a good substitute. The best varnish to use is the photographer's negative varnish, and this may be applied lightly with a soft brush. The varnish dries very rapidly, and therefore the glass should be covered as quickly as possible. It is better, too, to varnish the slides in front of a fire; for this not only accelerates the drying, but also tends to greater transparency.

It may be mentioned here that the photographer seldom applies his varnish by means of a brush, but flows it over the negative. This is a far better plan for lantern sketches also, for it never leads to the smearing, or other damaging, of the ink lines.

2. Pencil Sketches on Ground Glass

These are, perhaps, the easiest to make of all lantern slides, and they answer the purpose of the teacher admirably.

Some squares of finely-ground glass are required similar to that employed in the construction of the child's 'drawing slate,' and cut to the standard size as above described. The pencil sketches are then drawn (or traced) on these with a very hard, sharply-pointed pencil. The lines must be as black as possible, though fine; and as this necessitates a rather heavy pressure of the pencil, the point is rapidly worn away. Hence it is advisable to have a strip of fine emery cloth at hand, so that the point may be restored, with but little waste of time, after each few lines of the drawing.

One great advantage of this method of producing slides is the ease with which shadings of all depths can be obtained.

If the finished drawing be placed in the lantern, and the image thrown on the screen, it will be observed that the result is not by any means satisfactory. This is due to the dispersion of the light by means of the roughened surface of the glass, which dispersion, by reducing the light on the screen, renders the image very dull and indistinct. This may be remedied, however, by varnishing the slide as above recommended. The varnish, by filling up the little hollows in the ground surface, reduces the dispersion and, at the same time, it fixes the drawing so effectually that no cover glass is required, and even allows of the washing of the slide without damage to the sketch.

3. Slides made with Natural Objects

Many interesting natural objects are so thin that they may be mounted between two glasses to serve as lantern slides. Such objects, if really suitable, are by far preferable to drawings, or even photographs. If they are opaque, a mere shadow is projected on the screen, thus revealing nothing but the general form; but even this is sometimes useful, especially when the object used is one which does not reveal much variety of structure to the direct vision, and when the form is the special feature to which attention is to be given. Thus, if the subject in hand is a consideration of the various ways in which leaves are divided and the purposes served by these divisions, a number of pressed leaves, secured between glasses, will prove extremely useful.

Many natural objects, however, are sufficiently transparent to show clearly their internal structure when viewed in this way, and they thus produce the most perfect and beautiful illustrations possible. Among these we may mention, as examples, a portion of the skin of a fish, to show the structure and arrangement of the scales, the cast skin (slough) of the snake or lizard, small down or body feathers, the wings of insects, some delicate vegetable structures such as very thin leaves, fronds of ferns, moss plants, and some seaweeds.

Of course it is possible to render the majority of opaque animal and vegetable structures transparent, but the processes are very various, differing according to the nature of the objects concerned, and therefore beyond the province of this work. Those readers who are acquainted with the usual methods of preparing and mounting objects for the microscope will find no difficulty in carrying out the processes to which we refer, the only important difference being that the objects are mounted between glasses of 3 ¼ in. square instead of the smaller 'slips' and 'covers' used for microscopic work.

There are various other ways of producing satisfactory lantern slides, but the three we have mentioned will probably answer the purposes of most teachers. The reader has probably noticed that the disc of light thrown on the screen when a lantern slide is projected is usually circular or of some other desired shape. This form is given by placing a mask of black paper between the two glasses of the slide. Of course it is not essential, but it may be as well to mention that 'masks' of various shapes, to suit the different subjects of the slides, may be purchased at a low cost from the optician. Our own plan is to dispense with masks altogether as far as the slides themselves are concerned, whether the picture is glass-covered or not, but to glue on to the slide-carrier a mask of strong, opaque paper, with a clean, three-inch circle cut out, thus making the one mask answer for all the slides used.

Some optical lanterns are provided with a horizontal stage and an erecting prism. These appliances are very valuable for nature study work, since they allow one to throw on the screen images of any transparent object without the necessity of any mounting. Small tanks of water, containing living aquatic creatures (many of which are so transparent that all their internal organs are distinctly visible) may also be placed on the stage, and the structure and movements exhibited on the screen.

— 21 —

Nature Notebooks & Diaries

We have several times referred to the importance of nature notebooks and diaries in which the children can record their observations, and we now propose to give a few hints concerning them.

Each child should be provided with a rough notebook, of convenient size, for outdoor observations, whether in the school garden or in the field. This book will be used for sketches and brief memoranda which will afterwards be more neatly entered up at leisure. Seeing that the book is, therefore, only of a temporary nature, it may consist of simply a few sheets of paper folded together.

The permanent notebook in which all sketches and records are to be finally entered may consist of alternate leaves of ruled writing paper and drawing paper; but there is no reason why all the leaves should not be plain, for the space will probably be occupied principally by sketches, and the small amount of writing required may be done without the aid of ruled lines. Children have to learn to write without lines sooner or later. Why not sooner?

All lengthy descriptions of natural objects and phenomena will rather form part of the exercises in English, and these are more conveniently written in ruled books kept specially for the purpose. In the nature notebooks we should chiefly aim at descriptive sketches, with only as much written matter as is required to express those features which the sketches do not clearly show.

The nature diaries will vary according to the age of the children, and it is advisable to encourage the senior scholars to start a new book of this kind even if they have already been keeping one during their earlier years. If such a fresh start is made there is no reason why the more interesting and useful facts and observations should not be copied from the old book into the new.

In infants' classes the best diary is one that is large enough for the whole class to see, and which is filled in, day by day, by the teacher, the latter encouraging the children to offer suggestions for the entries. The diary may consist of a large sheet of brown paper, ruled up for the week, with a space for each day. Or it may even be ruled to last a whole month. On this the teacher might enter, by means of chalk, the state of the weather and any observations of value suggested by the scholars, more especially the things observed on the way to school, and any changes or developments connected with the plants or other specimens kept in the schoolroom or in the school garden.

When the children are old enough to keep a diary for themselves the teacher will decide on the most suitable form of book, and also on the manner in which the entries are to be made.

For the younger children a book of about forty pages is ample. Let them write the names of the months of the year at the top of each page, using an open folio of two pages for each month, and then enter their own observations under these headings. The remaining pages of the book may be used for miscellaneous observations and descriptions which are not necessarily connected with any particular month or season.

It will be seen that the simple notebook just described is not a diary in the strict sense of the term, though, of course, daily entries might be made if desired, in which case a much larger number of pages would be required if the book is to last a few years. The idea, at this stage of the child's education, is not to enforce frequent entries, but rather to see what the child takes a pleasure in recording. The teacher encourages, rather than forces the child, and leaves it as much as possible to its own initiative, giving occasional advice, and avoiding such rigidity of method as may tend to make the work a toil rather than a pleasure.

When the child is a few years older, and has reached one of the upper classes, it might be stimulated to commence a diary of a more useful and permanent nature. In this instance the book might consist of at least 150 pages, and be ruled more closely than is usual for ordinary written exercises.

The diary is at first prepared by fixing a space for each day of the year. Allow only a quarter of a page a day for the months of January, February, October, November, and December—there being naturally fewer observations to make during these months when Nature is more or less at rest, and when the weather is frequently less favorable for outdoor observations. For the remaining months half a page a day might be allowed.

The entries in this diary should be made as briefly as possible, the year of entry being indicated at the end of each one; and the remaining pages at the end of the book, not required for the daily records, could be utilized for fuller descriptions of special interest, together with accounts of the continuous observations made regarding the life histories of any living beings that have been watched through their various stages.

Here, again, the children should not be forced to make numerous entries. Let them have the fullest liberty to follow their own inclinations. They should not feel bound to make an entry every day, but simply encouraged to record those facts which have interested them, the teacher advising, but not commanding. Any interesting events recorded in the previous notebook of the child's earlier years, and which appear to be of permanent interest, might constitute the first entries of the new diary; but this, again, should be left to the child's discretion.

It will be seen that a nature diary, such as we recommend, may last several years, perhaps long after the school days are over; and it is probable that many of the old boys and girls, retaining the interest in Nature that was instilled during their school days, will start a new diary when the old one has no longer the space for further records.

Index

T

V

W

Y